Yearbook of the Artificial

Yearbook of the *Artificial* Volume 4

Edited by
Massimo Negrotti

In the different areas of the artificial – defined as the attempt to reproduce natural objects or processes by means of the available technology – researchers and designers often work within their disciplines with little or no information about research being carried out in other areas. Bioengineers, for instance, know little about the problems of Artificial Intelligence researchers, while roboticians often neglect the efforts of medical engineering, etc. The Yearbook of the Artificial discusses current theories, projects and models from a wide range of areas without going into too many technical details. The main assumption is that there are common logical and methodological problems for all disciplines that are concerned with the reproduction of natural objects and processes. The Yearbook provides a forum for all of them, and covers areas such as bioengineering, robotics, A.I., artificial neural networks, artificial life, multimedia, the history of technology, communications, art and music.

In submitting a paper to the *Yearbook of the Artificial*, you grant us permission to publish that paper after a peer review. We ask that your paper be between 20 000 and 24000 characters, and pertinent to the topics mentioned above. We also require an abstract of no more than 150 words, including a list of keywords. Papers should be submitted in triplicate, typewritten, double-spaced, on one side of the paper. Please include a copy on a floppy disk in a commonly used word processor format.

Only original papers in English should be sent to:

Dr. Domenico Donnanno
dodo@soc.uniurb.it
IMES-LCA
Via Saffi, 15
University of Urbino
I-61029 Urbino (Pu) Italy

Yearbook of the Artificial

Nature, Culture & Technology

Vol. 4

Kyosei, Culture and Sustainable Technology

Edited by
Massimo Negrotti &
Fumihiko Satofuka

PETER LANG
Bern Berlin Bruxelles Frankfurt am Main New York Oxford Wien

Bibliographical information published by Die Deutsche Bibliothek
Die Deutsche Bibliothek lists this publication in the Deutsche Nationalbibliografie; detailed bibliographical information is available on the Internet at ‹http://dnb.ddb.de›.

British Library and Library of Congress Cataloguing-in-Publication Data:
A catalogue record for this book is available from *The British Library,* Great Britain, and from *The Library of Congress,* USA.

Il presente volume è pubblicato nell'ambito della Ricerca 'La cultura dell'utente. Cittadinanza, mercato, tecnologia', finanziata dal M.I.U.R. e dall'Università degli Studi di Urbino 'Carlo Bo', PRIN 2004, Prot. n. 2004149922.

This volume is published as part of the research project 'The culture of the User. Citizenship, Market, Technology', co-financed by the Italian Ministry of University Teaching and Scientific Research (M.I.U.R.) and by the University of Urbino 'Carlo Bo', PRIN 2004, Prot. n. 2004149922.

Cover image: Massimo Negrotti
Cover design: Thomas Jaberg, Peter Lang AG

ISSN 1660-1084
ISBN 3-03910-972-3
US-ISBN 0-8204-8363-X

Hochfeldstrasse 32, Postfach 746, CH-3000 Bern 9, Switzerland
info@peterlang.com, www.peterlang.com, www.peterlang.net

Scientific Committee

Table of Contents

Massimo Negrotti

Foreword

Due to the large number of papers contributed to the third volume of the *Yearbook of the Artificial*, a large part of the fourth volume once again concerns the theme of the 'user' of the artificial, which was the underlying theme of Volume III. Nevertheless, we have tried to gather in Volume IV papers dedicated to three almost coherent areas. The first involves the discussion of some aspects of the theoretical basis upon which we may interpret, from different standpoints, the relationships between technological artifacts and human beings, thanks to papers by Pasi Falk, Karamjit S. Gill, Stefano A. E. Leoni, Victor Margolin, and Giuseppe O. Longo. The second regards applied studies of the user in different fields, ranging from personal identities, to the Internet, to planning, to special case studies in Italian society, thanks to contributions by Mariella Combi, Renata Livraghi, Giuseppe Padovani, Antonino Porrello, Barbara Rondelli, and Massimo Russo. Finally, the third area, introduced separately by Fumihiko Satofuka, refers to the concept of *Kyosei* (symbiosis) as a key point for the study of contemporary technology-related social phenomena.

Massimo Negrotti, LCA – Laboratory for the Culture of the Artificial, University of Urbino

Mariella Combi

The construction of flexible identities and negotiation of meanings on the Internet

Abstract:

In mixed cultural situations with migrations of people and meanings we observe a fluidity of cultures: it is necessary not only to know how languages communicate but also how words carry meanings that contribute to the buildup of personal and community identity, and solicit new representations and new images of self. The case of 'multicultural users'.

Keywords:
multicultural user, flexible identities, web communication, cultural anthropology

1 Introduction

Anthropology today confronts a major challenge in dealing with cosmopolitan cultural forms and the occidental experience of them, notably in the model concerning world computerisation.

The theoretical scope of the anthropology of our time is, therefore, the context within which I analyze the role of communication when different cultural models are met in the virtual place of computer technologies, i. e. Internet. The fact that writing has so far prevailed makes it possible to know the linguistic choices that may influence another's perception and one's own, together with the construction of identities I call 'flexible'. These allow a single person to express almost simultaneously a variety of cultural belongings. Such is the case of immigrants in their relationship with the West, with other people coming from different geographical and cultural areas, with those who are encountered virtually.

The social, territorial and cultural changes that have occurred in today's world, in the reproduction of the sense of belonging to a group or groups and, even earlier, in questioning the concept of cultural identity marks the limits of its thematic frame.

In anthropological research the question arises from the impressive displacement of people taking place throughout the world, with the result that migrating groups settle down in new localities and set up new cultural 'projects'. It follows that, on one hand, the subject/object of anthropological research is no longer represented by groups limited territorially or culturally homogeneous. On the other hand, new virtual communities appear on the scene.

Dwelling on these topics leads not only to remarks about the use we make of cultural representation, but also to the complex process of imaginative appropriation implicit in the fluent building up of linguistic communication, both in geographical space and, by means of the computer, in virtual time.

Communication is fundamental to the anthropological definition of cultural context and the sense of perceived belonging. The process of identity building requires communication in its technical, cultural and social dimensions, and the attribution of a role by the use of new terms may provoke the re-elaboration of conventional models.

The anthropologist's interest is attracted by relationships which have always been the centre of his research: the connections existing between culture and communication and the role of language

i) in negotiating for meanings that allow virtual, multicultural connections and
ii) in constructing capabilities for the simultaneous control of symbolic structures.

Every society organizes always its knowledge so that it may be communicated, and thus transferred to build up the sense of belonging called 'cultural identity'. And this is obtained through rules and symbols drawing world representations which give sense to reality: from stories to lines of conduct, from language to actions, from emotions to their representation, from concreteness to virtuality. One belongs to a community first of all through the knowledge of its language: that is why the anthropological dimension of communication always refers to a world vision.

Technology is, in fact, a complex cultural system in which anthropological and linguistic components are inextricably involved and have an influence on its meaning and social practices. For this reason, actions that include verbal and written communications are the expression of the social structure through which use of language in everyday life can be analyzed. Thus a culture can also be defined as a set of learned behavioural models and interpretative practices where language is the essential system bringing order to experience. I consider language as a set of cultural practices in which the relationships linking word and world, self-perception and belonging are analyzed.

In mixed cultural situations with migrations of people and meanings we observe a fluidity of cultures: it is necessary not only to know how languages communicate but also how words carry meanings that contribute to the buildup of personal and community identity, and solicit new representations and new images of self.

Technological change has made the imaginary a collective fact fundamental to everyday experience, all the more so because our perception of technology is in constant movement between the imagination and concrete activity. This movement has an incidence on the characteristics of technology by inserting them into the everyday experience of the individual. Verbal communication, the negotiation of meanings, and the construction of flexible identity come into play here.

The attention of the anthropologist must be directed to the way signs are used:

i) in the daily displacement of people during migration, and
ii) in the use of Internet where meaning is displaced by virtual communication without the displacement of the users themselves.

2 The case of 'multicultural users'

Massimo Negrotti has co-ordinated a social national research project on 'The models of user in advanced societies', part of which is the text entitled '*The multicultural user and new technologies. A cultural anthropological approach*'. This text is designed to illustrate aspects of anthropological research on web communication whose objective, as I proposed it for the research unit, was to create an anthropological model of a 'multicultural user' defined by analysis of this term in official documents on the Internet[1].

In today's world the buildup of cultural realities without frontiers and 'global' technological culture of occidental origin may work to influence the realisation of new social bonds, both formal and informal, as well as causing crises of 'place' and local culture.

However, in the course of intense data surveying activity a difficulty arose: the phrase 'multicultural user' appeared nowhere in the web sites and documents of the Italian ministries of Health and Education. But the concept of the multicultural user seemed to be confirmed by the frequent occurrence in many different sectors of the term user. The idea of multiculturalism was, for our purposes, a specific term limiting the dispersion of the word 'user' by establishing a direct connection between the world of immigration and its cultural representations in contact with public administration and ordinary everyday life. This unexpected result invites reflection on the reasons for that absence and its potential for communication: it places the role of the person in the social order in doubt. Indeed, this role and its image are also established, very often unconsciously, by the choice of words in documents, laws, decrees etc. These terms are not only instruments essential to our knowledge of the world and the creation of relationships, they also define roles, values, rights, duties in particular with institutions.

Thus far the model 'user', as distinct from 'foreign user', is reasonably well supported by analysis of documents of Italian (Health; Education) institutions. The meaning of the word 'user' has now to be

1 The research unit is at the University 'La Sapienza' in Rome. Among the members of the team were researchers Gabriella Corona and Maria Grazia Corradini.

analyzed from the point of view of the people to whom these documents are directed. To this end it becomes useful to undertake ethnographical research which goes forward from the hypothesis that Italians and Italian-speaking cultural mediators and immigrants have different perceptions of themselves as members of a community if the language of official documents and web sites refers to them as '*user*', '*person*', '*foreigner*' or '*citizen*'. Is it possible to assert that the more one is regarded as a user, the less one is considered a person? How do the terms 'user' or 'foreigner' configure their presence in a community space, among the other citizens?

The notion '*multicultural user*' reminds us that the foreigner enjoys the legitimacy of Italians themselves. Thus the image of a collectivity as an undistinguishable ensemble of individuals is perceived as less emotionally involving, because quantity nullifies the singularity that disturbs.

The task for the anthropologist is to find out how to perceive the social and political position of immigrants and cultural mediators, especially that of women, by means of a field research. These subjects will talk about their personal, everyday experience as *users*, *citizens*, *people*, *patients*, *customers* of institutions, and in so doing reveal the characteristics of a possible cultural negotiation.

In anthropology, in order for an individual to be recognised as a *person*, a symbolic order and shared cultural representations must confer upon him a place and a role in society beyond legal and moral acknowledgement.

The innocuous remark 'We are all multicultural' reminds us that we are all part of a multicultural entity and that each of use represent one of its aspects, in particular.

To reiterate to the main question: the assignment is to find out how the absence from official documents already analyzed of the term 'multicultural user' influences the models of social interaction between institutions, Italians, and immigrants in Italy.

Bibliography

A. Appadurai, 'Global Ethnoscapes. Notes and Queries for a Transnational Anthropology', in R. G. Fox (ed.), *Recapturing Anthropology*, School of American Research Press, Santa Fe 1991.

A. Appadurai, *Modernity at Large. Cultural Dimensions of Globalization*, University of Minnesota Press, Minneapolis-London 1996.

M. Combi, 'The cultural body and artificial technology', in M. Negrotti (ed.), *Yearbook of the Artificial*, Vol. 1, Peter Lang, Bern-Berlin 2002.

M. Combi, 'Cultural Models in Anthropology', in M. Negrotti (ed.), *Yearbook of the Artificial*, Vol. 2, Peter Lang, Bern-Berlin 2003.

M. Featherstone (a cura di), *La cultura dislocata. Globalizzazione, postmodernismo, identità*, SEAM, Roma 1998.

U. Hannertz, *La diversità culturale*, Il Mulino, Bologna 2001.

M. Herzfeld, *Cultural Intimacy. Social poetics in the Nation-State*, Routledge, New York 1997.

M. Negrotti, *Theory of the Artificial*, Intellect Books, Exeter 1999.

M. Negrotti, *Naturoids. On the Nature of the Artificial*, World Scientific Publishing, Singapore 2002.

M. Negrotti, 'Parliamo, ascoltiamo, comunichiamo ma ancora continuiamo a non capirci', in Telèma, n. 24, 2001.

D. Wolton, *Internet et après? Une theorie critique des nouveaux medias*, Flammarion, Paris 2000.

Maria Domenica Combi, Dipartimento di Studi romanzi, Università 'La Sapienza', Roma

Pasi Falk

Universe as an artifact: from God's creation to digital metaphysics

> *In the beginning was the Word, and the Word was with God, and the Word was God*
>
> St. John, *Gospel, 1.1*

Abstract:

So, God was the first creative artisan, in other words, the supreme architect and artist creating the world and thus the primus motor *of artificiality – that is, if we stretch the term to include not only human but superhuman or divine actors. Actually the human culture – and its mythical projection into cosmic scale – was from the very start an artifact, long before the Judeo-Christian monotheism and its 'logocentrism'. However, the cosmogonic formula presented by St. John is specifically symptomatic and illustrative while referring to the other of the two originary artifacts of human culture: the 'word' and language as a dynamic structure of words spoken and heard, written and read. The other originary artifact – considered by the more technologically oriented observer as the first and primary one – relates to the creativity of man as 'tool making animal'. However, the broader scope of the anthropological studies of material culture regards the symbolic (and even artistic) dimension of artisanship as important as the functional tool making – and, as it appears, with good reasons: a more deep going analysis of these two primal scenes reveals that they are actually two inseparable aspects of the one and same scene (as will be demonstrated in the paper).*

In the dawn of modern (Western) natural science, especially from the 17th century onwards (since Descartes, Leibniz and Newton constitutive discourses), the universe – and thus the nature in the widest sense of the term – was still conceived of as a divine artifact, the creation of God but now as an artifact separated from its creator and granted an autonomous existence as an object of scientific inquiry: the 'book of nature' should be studied with rational means, analytic method (Descartes) and logico-mathematical means – without reference to the other book, the Holy Bible. The universe should be seen as a mechanism or machine (Descartes) and logical entity (Leibniz) obeying the laws of nature accessible for the rational mind with logico-

mathematical means. Thus the secrets of the universe and nature would be revealed by analyzing it down into its simplest parts – both conceptually and physically – and thus finding out how it was/is constructed and how it works.

This implied a dramatic break with the taxonomic-hierarchical theo-cosmologic scheme – the 'great chain of being' (Lovejoy) – which had a substantial influence on the 'natural history' tradition preceding natural science but nevertheless maintaining its status as a legitimate approach in studies concerning the (esp. organic) terrestrial nature till its 'scientification' as evolutionary theory during the 19th century. This is the very constellation which gave rise to the romantic reaction against the scientification, objectification and economic utilization of the nature motivated by a nostalgic longing for the lost unity of man and nature (within confines of the great chain of being) and actually 'inventing' the pure and genuine nature as a utopist counter model for the rational enlightenment. This reaction was accompanied with a revival of natural philosophy – a tradition of thought that has survived till today in a number of eco-philosophical variations.

However, something curious is happening in the recent phase of the digital era. In a certain sense, the flag ship of the enlightenment, the natural science, seems to be entering the final harbor of complete knowledge of the book of nature. Nevertheless, at the same time it is at risk of turning its object into pure abstraction, or, in the worst case, losing its object(s) altogether. The three main fields of natural science – biology, chemistry and physics (plus all the derivative sub fields) – are integrated into one complex with a common denominator called 'information'. Today, these fields of research are not only heavily 'computer aided' but thoroughly 'informatized': bio-informatics, genomics, proteomics etc. deals with the informational processes at the molecular level; neuro-informatics, cognitive science and artificial intelligence at the neural(net) level; quantum informatics and digital physics at the atomic and sub-atomic level – and finally nano-informatics, nanoscience and nanotechnology operates between all these previous levels with an aim for technological innovations.

Surely my aim is not (in the paper) to deconstruct the contemporary technoscience complex along all the various relevant dimensions but merely to thematize the way it redefines the concept of nature – in concepts (and research fields) such as 'artificial life' and 'digital physics', the latter of which ends up in the search for the 'ultimate constituents' of matter into pure information, pure differences, zero's and one's, as it were [f.ex. Wheeler, Fredkin and Deutsch]. The irony of these interpretations is that it brings back the Creator – not into the theoretical explanations but – as a bracketed background figure. Namely, if our universe is ultimately pure information processing and thus a computer program then a 'natural' question would be: who's the programmer, the supreme Hacker? No wonder some religious circles have today great expectations concerning the recent developments of quantum physics...

Keywords:
difference, cosmology, cosmogony, physics, technoscience, digital, informatics, information

1 Introduction

My specific focus will be on *the genealogy of 'nature' as an epistemic object* – and thus I am dealing with 'nature' in the widest sense of the term which posits it in close range with terms like 'world' and 'universe'. Certainly I am not going present you condensed version of the whole history of natural science. I will rather *thematize certain dimensions of the genealogy of nature oriented 'scientific curiosity'* which characterizes specifically Western (Judeo-Christian-Islamic & Greek-philosophical) tradition – the tradition celebrating today its newest phase named *digital era*, *information age* and so forth.

The main focus in this condensed genealogy of scientific or 'theoretical curiosity' [*theoretische Neugierde*; Blumenberg, 1984] will be on a specific dimension, a *constructivist*[1] *thinking* dominating this genealogy, especially from the 17th century onwards. That is, from the era witnessing the formulation of the bedrock of modern natural science: specifically Descartes' mechanistic universe and the 'method of analysis', Leibniz's mathematical innovations – including differential calculus and the 2-base number system – and their application to the epistemic object of (physical) 'nature' and the Newtonian project of classical (mechanical) physics. This genealogy does not proceed up to the present digital age as a smooth progressive continuum. There are many turns and (Kuhnian) ruptures concerning the epistemic object of 'nature' in all its scales, from cosmologic models to terrestrial (organic) life world and further down to the ever smaller building blocks of material reality.

My argument could be formulated as follows: *the contemporary digital technoscience complex pushes the constructivist orientation of*

1 I am using consciously this term in distinction to 'constructionism' which at the present is strongly associated to the problematic notion of 'social constructionism' [Hacking, 1999].

scientific curiosity – and all the systematic discursive (e.g. theories) and non-discursive (e.g. experiments) formations it involves and produces – *to the extreme, to the ultimate level.*

2 Digital technosciences

However, before I start explicating my argument I probably should outline the 'technoscience complex' as I understand it and say few words about the broader frame within which I am practicing my own scientific curiosity. So, to make a long story short, my curiosity is directed *to the role of digital technology and – more broadly – digital thinking in the production of scientific knowledge.* Consequently, this specifies the object of analysis into those sciences and fields of research which are 'digitally intensive', that is, to the integrated realm of contemporary technosciences. And in the present context I will – naturally, as it were – remain within the realm of natural sciences which actually form the (hard)core of the digital technoscience complex – presented schematically as follows (Fig. 1).

In the middle row we have the three basic disciplines of the natural science – biology, chemistry and physics – plus the more specific field of neurosciences due to its close relation to the digital dimension which dominates the whole technosciene complex. After all cognitive science took off in the 1950s by modeling the brain as an computer and the AI (artificial intelligence) aims to humanize the computer on the same grounds. Interestingly enough even the late 19th century neurology was heavily indebted to the discourse on electricity of the era creating a kind of symmetrical metaphoric: the nerve system was pictured as the telegraph network of the body – telegraph being the first substantial technical application of electricity – and the telegraph network as the nerve system of the society. As it happens both were discovered/invented approximately at the same time (in the turn of 18th and 19th centuries) thus giving good grounds for the symmetrical use of the metaphor – between the natural and the artificial, as it were.

The links between the four fields of (natural) science have been growing together since the end of the 19th century, on the one hand, due to the advances in the intermediate areas of research, and, on the other hand, due to the mathematization of these fields, starting from physics already in the 17th century (Descartes and Leibniz) and then moving gradually left till covering the entire row (and invading some other non-natural disciplines excluded from the scheme as well). However, the primary logico-mathematical integration of the technosciences is realized in a certain form, that is, as computation and all the algorithmic elaboration it involves, thus justifying the label 'digital' as well as its kin term 'information'. This integrative dimension is located in the box above acting as a theoretical umbrella for the four fields.

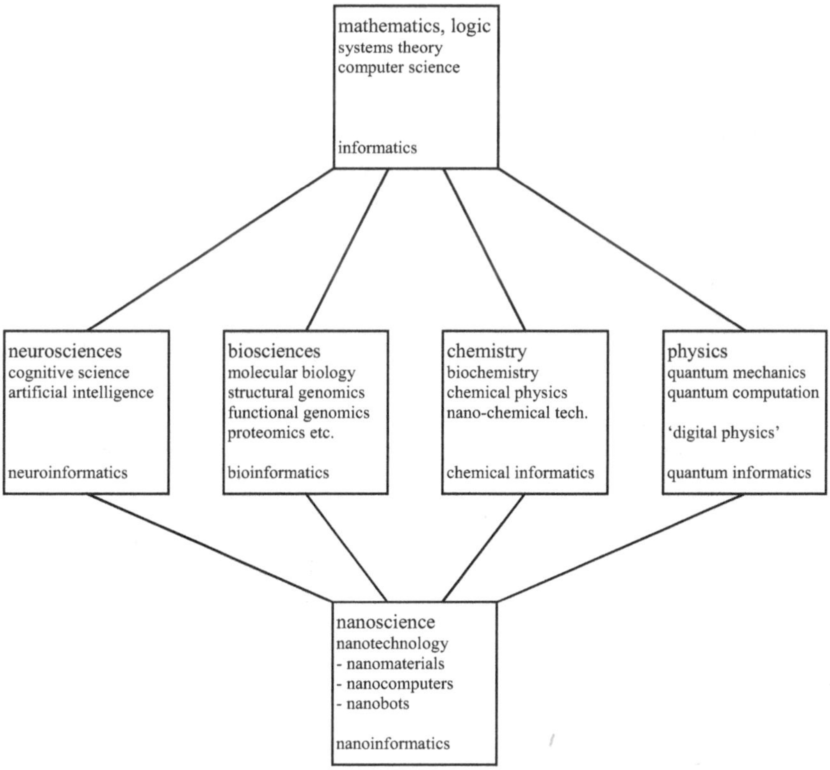

Fig. 1. Digital technoscience complex

As the scheme indicates, the common denominator of all six modules is *informatics*[2] (or information theory/science), a term which in the contemporary technoscience complex gains a meaning extending far beyond the sense it has as part of 'library science' (defining information at a higher level organization as knowledge; cf. also the net-updated version called 'data mining'). Surely, the term still refers to gathering, storing, retrieving and classifying recorded information but now it includes more elaborate and complex modes of information manipulation and processing as well. And furthermore, it now covers the whole scale of complexity levels down to the elementary binary basic level of informational 'atoms', that is, the bits.

Consequently, the boundaries between informatics and the other fields in this module – from computer science via systems theory to mathematics – seems to fade away. However, this does not mean that all these fields would operate with a shared notion of information. As it appears, there are different understandings of information both within certain fields and especially between different fields [Chmielecki, 1998], ranging from the Shannonesque sense – which is purely mathematical and technical [Shannon, 1948] – to conceptions which approximate the linguistic and semiotic senses of signification and communication. One the other hand, one central source for these conceptual ambiguities seems to reside in the metaphoric formulations dominated by the 'computer' and its metonymic extensions downward (in scale) to the binary zero/one-foundation of 'information' and upward to 'network'.

So, the actual entities these disciplines, especially the (above) technosciences, are working with is not (in/organic) matter or energy but information. Certainly, the stuff as matter and energy is still there but the actual entities scrutinized are not defined in these material terms nor as abstractions of these thing-like objects but as structural

2 The combinatory term x+informatics is not necessarily very commonly used in all mentioned research fields – with the exception of bioinformatics – but this does not change the fact that all these fields are heavily dependent not only on digital information processing technology but also on 'digital thinking' (as my argument goes). This dependency is much stronger than the notions 'computer aided/assisted research' suggests.

relations (cf. structural genomics) and dynamic *processes* (cf. functional genomics) – processes that are rather specified as *translation, encoding* and *decoding* (cf. transcriptomics) or even *communication* than mere *transmission* of energy/matter or transformations within or between these two basic modes of physical existence.

3 Pure difference

However, refocusing the epistemic object from things to (structural) *relations* and (dynamic) processes is not an actual novelty. It was already part of the mathematized mechanistic cosmology and physics in late 17th century, especially from Newton onwards who refocused the epistemic object from 'things' (particles) to actions, motion and change. The digital[3] novelty lies in *the reduction of the relationship to its most elementary and abstract form – into pure difference*, into the basic zero/one-foundation of the binary or 2-base number system outlined already by Leibniz in late 17th century. It should be noted that neither zero or one is the basic unit. The basic unit is the pure difference between these two, or some other non-identicals (entities, states, etc.) So, the binary principle of pure difference is, on the one hand, a *categorical abstraction* which can be applied to whatever material and energetic entities – as long as these can be translated into complexes built of and functioning on differences.

On the other hand, the pure difference is the *ultimate basic unit of any construction*. It is an abstraction as a difference between two entities, be they abstract or not. Furthermore, as the basic level of any logico-mathematical operation it is a double abstraction: all these operations can be translated into those of the 2-base number system[4]. However, as it appears, the applicability of pure difference escalates beyond the mathematical boundaries first into a basic epistemological

3 Actually 'digital' refers simply to number (digit) and goes etymologically back 'finger' (lat. *digitus*), to counting with fingers (and toes). Today it refers primarily to the binary computational principle.

4 George Spencer Brown [1972] developed in the 1970s an specific logical system, a 'calculus of indications' which takes off from pure difference.

principle and then to the ontological realm and finally to metaphysics. After all, the existence of (pure) difference, being different and being (or becoming) amounts at this basic level to the same 'thing', as a pure relatedness.

Thus the binary principle is turned into a universal foundation and starting point for construction as such or better for the construction of anything of whatever stuff, as long as there are differences, however constituted. Nevertheless, one (two-part) condition must be fulfilled for reaching necessary consistency: the entity or the construction site should be *finite* – a whole with boundaries – and consist of *discrete* parts, whatever their mode of existence or scale might be. Discreteness is the necessary & sufficient condition for difference and finity (or limitedness) the necessary & sufficient condition for construction. Otherwise the construction – the universe, for example Descartes' machine, would never be completed (into a whole) nor comprehended (as a structured/dynamic whole)[5].

This could be understood as the boundary between physics (& physical ontology) and metaphysics. The latter deals, in one way or another, with the absolute notions (givens) of time and space as eternity and infinity and void[6]. For construction – and thus for modeling[7] as well – these notions are simply useless, except in mathematical formulations in which they are subsumed under the rules and within the confines of finite world(s), as limits (Ø & ∞) thus losing thus their absoluteness. This is the role of these notions for example in differential calculus, outlined already by Leibniz and Newton (separately) in

5 Note the two senses of the term 'comprehension': 'understanding' and 'making something comprehensive', into a (all)inclusive whole'.

6 These was no void in Descartes' universe: extension was characteristic only to the material world (*res extensa*). At the present the void seems to be disappearing again, due to notions such as dark energy and anti-matter. The 'ether' – the old stuff of the void – is making a comeback.

7 Modeling is a work of construction in reverse, to use an architectural metaphor: the architect/constructor elaborates a blueprint then realized in the actual construction while the modeler proceeds in the opposite direction aiming at discovering the 'blueprint' (the dynamics, laws of nature etc.) of the given 'construction' – given or designed my the divine architect (Descartes, Leibniz) or simply given (after the 'death of God').

17th century. So, posing questions about what lies beyond the boundaries of the whole/construction, or what preceded the 'big bang' etc. is doomed to end up in an eternal return of the same – barren question. The only way out of this vicious circle is to set up the boundaries; to construct, to model, to imagine or believe. This is the common ground for both to scientific (physical) cosmologies and (religious) myths. In fact the setting up or marking of boundaries and the representation of what could be called the *originary difference* is the basic formula of any mythical cosmology & cosmogony [Sahlins, 1983; Siikala, 1991; Valeri, 1985] – tempting me into a short (deconstructive) excursus on the logocentric Judeo-Christian version of cosmogony.

4 Excursus: originary difference

As known, St. John's gospel (1.1) starts as follows: 'In the beginning was the Word, and the Word *was with God*, and the *Word was God…*' (emphasis added). This should suffice for a minor deconstructive exercise, that is, to look more closely at the contradictory relationship between these two sentences and trace its unsolvable 'supplementary' character [Derrida, 1981]. So, is God the unity before the originary difference or the void before the creation (ex nihilo). Relating to Leibniz's theological interpretation of the (pure) two-base difference this would turn God into void (zero) which is just the opposite Leibniz proposes, For him one (1) stood for God and zero (Ø) for void. But God is the Creator and there was nothing before the act of creation – except for the Word which was 'with God' and actually 'was God'. Consequently God is both the agent of creation and the act of creation itself, or, both the precondition of the act and the act itself[8] – and finally the agent in charge of the creation (that is, the outcome of the act). The last role is bracketed already by Descartes while the two former are pushed outside the epistemic object of nature/universe in later times.

8 No wonder St. Augustine tried to solve in his 'Confessions' this unsolvable configuration by replacing the preceding state of 'nothing' (*nihil*) with 'almost nothing' [*prope nihil*, see Kaplan, 1999, p. 280].

Interestingly enough, the same supplementarity figures in the etymological roots of Greek 'chaos' as well. Its original sense is that of rent or a 'violent tear' [Steiner, 2002, p. 29] or *splitting* referring thus to the act of creation or to the moment of the originary difference. However, the role and sense of the term is transformed in the dichotomy of chaos vs. cosmos where chaos refers to the state of disorder preceding the act of creation/becoming which brings forth *order*. This is reminiscent of entropy vs. negentropy distinction of the 2nd principle of thermodynamics first formulated in mid 19th century and linked closely to the concept of information in the next century (esp. Shannon's formulation: information = the degree of negentropy, in relation to the entropic 'noise'), [Shannon, 1948][9].

Furthermore, traces of the same supplementarity is found in the (translation) etymology of 'nature' as well. As Tellenbach and Kimura notes

> the Latin *natura* [...] stems from *gnascor* which means to be born, grow, emerge, originate. *Gnascor* is related to the synonymous Greek term *gignomai*. Thus nature is really the translation of genesis. The Romans, however, employed *natura* to translate the Greek term *physis* ... [Tellenbach and Kimura, 1979, p. 177].

However, the translation error is not too severe in case one draws on Heidegger's explication of the Greek term 'physis' as

> that which comes forth of itself (e.g. the sprouting of a rose), that which in opening itself up unfolds, and in unfolding makes its appearance, maintains itself and abides – in short, that which in coming forth and abiding, governs. [Heidegger, 1953, p. 11].

So, physis is both being and becoming or 'coming into being'.

9 In contemporary dynamic systems theory, focusing on the self-organizing dynamics (emergence, autopoiesis), the emphasis has shifted from the deterministic entropic fate to the negentropic processes where chaos is redefined from an indifferent absolute disorder into a creative condition producing order [see Capra, 1997, Prigogine and Stengers, 1982, pp. 135-155, Prigogine and Stengers, 1985].

5 Parts and whole

In order to explicate more clearly the 'coming into being' of the pure difference we have to link it to another distinction, that of *parts vs. whole*[10]. This distiction has a central role in the genealogy of the (Western) nature oriented thinking. The link implies a specification of the 'scientific curiosity' in terms of scale. If it is natural (as it were) to direct human curiosity horizontally, at the human (body & perception) meso-scale, towards the surrounding world and if the logic of myths makes the religious/scientific curiosity towards cosmic dimensions conceivable, then why would curiosity be oriented down the scale towards the bottom? Explaining the 'will to knowledge'[11] as a unversal human disposition[12] does not work. From a historical point of view cultures differ considerably in this respect and historical change of a certain culture makes the difference as well, as in the case of the Western tradition: in Greek antiquity there was more degrees of freedom for such curiosity than during the later centuries when the church categorized curiosity among other vices. So, for example,

So, for example, the ideas of the pre-Socratic atomists [Heraclitus, Epicurus *et al.*] were buried for a long time and were truly resurrected not until the 19th century. Nevertheless, epistemic orientation towards the bottom of things were already there and it was actually systematized into a 'scientific' principle in Descartes 'method of analysis'[13] in

10 The topic has produced even a logico-formal discipline of its own named 'Mereology' or simply 'ontology' by the founding figure Stanislaw Lésniewski in the 1920s. As Simons notes, this was 'the first extensional part-whole theory to be rigorously developed' [Simons, 1987].

11 Perhaps comparable with both Foucault's 'will to truth' (*la volonté de savoir*) and Nietzsche's 'unscrupulous curiosity' (*rücksichtlose Neugierde*) in 'Beyond Good and Evil' (1980 [1886], § 180, p. 646) even though Foucault's concept derives primarily from Nietzsche's 'will to power' (*Wille zur Macht*).

12 The classical note on such universal human disposition is found in the opening words of Aristotle's *Metaphysics* (c. 350 B.C.): 'all men by nature desire to know...'.

13 The 'Discourse on Method' was published 1637 and the more detailed exploration of the philosophical basis for his new approach to knowledge appeared a bit later in his 'Meditations on First Philosophy' (1641). His 'Rules for the direction of the

the 17th century. These were the times when curiosity – especially the scientific form – gradually turned from vice to virtue[14].

Descartes' method is presented as a general guideline for systematic scientific thinking, as a 'method for investigating the truth about things', as he notes in rule IV (Rules). That is, its application was not restricted to the investigation of nature or the physical world. However, the principle is closely related to an atomist conceptions and it may well be argued that Descartes analytical method was modeled on these grounds. Actually, he studied intensively Isaac Beeckman's 'micro-level mechanical explanations of natural phenomena', which could by and large be characterized as atomist, some 10 years before 'Rules' [Dyer, 2001, p. 81]. So, even though Descartes' notion of 'analysis' is located primarily in the toolbox of scientific thinking – as an abstraction[15] – it nevertheless refers to more concrete (material/physical) sense of the term as well, such as the concept of analysis in conventional chemistry: taking the compound apart in separate elements. The principle of distinguishing 'what is most simple from what is complex…' (rule VI) applies here as well. Consequently, it is not enough to contextualize Descartes method only to the long discursive history of the *analysis-synthesis relationship* even though Descartes' distinctive approach stands out in this context as well; [see Timmermans, 1999]. Descartes' 'analysis' should be paired with construction as well because it not only presupposes a construed whole and a 'constructivist stance' but is actually a project of construction – in reverse.

What could this mean? Why would not analysis be a genuine and perhaps primary orientation of curiosity, scientific or not? Is it not

---

mind'(Rules) was written in 1628-1629 but not published until 1701 (all these texts are in the net).

14 According to Samuel Johnson ('Dictionary', 1755) 'to be curious' or 'to be addicted to inquiry' is justified and even virtuous in contrast to the negatively charged mode of 'additction, like the impassinated desire for novel commodity, or extreme instance of passionate attachment [which entails a] complete or partial surrender of self-government before an external object or agent'.

15 Rule XIII: 'If we are to understand a *problem* perfectly, we must free it from any superfluous *conceptions*, reduce it to the simplest *terms*, and by process of enumeration, split it into the smallest possible parts' (Rules/emphasis added).

natural (as it were) to look for the essence behind and beneath the appearance – to penetrate through the obvious and visible into to the hidden core of matter(s), from the known to the unknown or not yet known truths? Yes and no. This has been the dominant principle in the Western logocentric orientation but it has never existed without the opposite pole, the holistic stance according to which the essence, truth and so forth is revealed not until having a perspective covering the 'big picture' – opening an understanding to the role of the parts. Such holism is not characteristic only of the other cultures, for example those of the Orient, but has been an constant element in the Western tradition as well, even though in a rather secondary position.

Nevertheless, from a constructive point of view the analytic 'taking the whole apart' principle is simply the only reasonable strategy: if you want to find out what a strange whole or construction is about you have to open it up, se how it is built and (if it includes motion) how it works. Actually, a disposition for such basic practical logic derives already from the pre-historical era of tool making, the primal scene of the (human) artifact: one of the two basic sources for the conception of whole/part is the making of a tool or some other artifact out of at least two different parts, like an axe with a stone head and wooden shaft. The other source is language, however primitive, and the hierarchical taxonomy as a necessarily structure in it[16].

The same analytic-constructivist 'taking the whole apart' principle still applies when going down the scale, even when this is done only in thought, imagination and speculation, as in the case of pre-Socratic atomists: taking partial wholes apart and then the next ones etc. until reaching the homogenous ultimate building blocks. Nevertheless, it

16 Actually tool making and language are closely linked together: language is born in the act of distinction (difference coming into being). It corresponds to the originary difference of the mythical cosmogonies. Both begins with the originary difference and proliferates into a hierarchical structure of differences: two aspects of the same process. In many less abstracted (so called primitive) languages the notion of 'everything' actually refers to the originary difference – structured in binary terms as Lévi-Strauss has explicated. For the *Hanunóo* in Southern Philippines 'everything' or universe is the sum (things) *nameable* + *unnameable* [Lévi-Strauss, 1996 [1962], p. 138]. For the *Hua* in Papua, New Guinea 'everything' is the sum *edible* + *inedible* [Meigs, 1988, pp. 341-357].

remains a constructivist stance aptly expressed in atomist questions like: what is the world *made* of? Followed by another question which was elevated into primary position with the mechanistic conceptions of nature and universe: how does *it* work? The 'it' – the world, nature or universe – must still be a coherent whole, a construction which, when put in motion, turns into a 'machine'. After all, the concept of 'part' necessarily presupposes – implicitly or explicitly – the 'whole', of which it is a 'part' [cf. Luhmann, 1988, p. 52]. And when moving towards the bottom the part is turned into the ultimate building brick or constituent of which the whole – God's mind or his creation, the machine or the universe modelled as a finite and thus thinkable or comprehensible – is constructed. This is the basis of classical physics and also of modern physics (after Einstein's general relativity) till the quantum mechanics which made things complicated – almost literally.

6 The basic formula – and its decline

So, the basic idea of science, scientific method and explanation – being through and through constructivist – may be formulated as follows:

whole :: part = complex :: simple

In other words the whole-part relationship corresponds to the complex-simple relationship. Thus the only scientific approach for understanding and explaining the whole is to go down to the simple parts and then back, step by step, to the complex whole, now explicated into a known structure and/or dynamics. And everything looked fine in physics till passing the critical boundary down the scale to sub-atomic levels which constituted the epistemic object of quantum mechanics. What was found was not simple but complex, or better complicated[17]

17 In systems theory a distinction is made between these two terms. 'Complexity' is a systemic and thus ordered quality while 'complicatedness' is a non-systemic quality.

so the basic formula was in crisis and is, in a sense, still today[18]. The term 'quantum' (originated by Max Planck and Albert Einstein) expresses aptly the newly found confusing complicatedness at the bottom: it is and it is not about particles or waves or energy so better call them just quanta. The semantics and metaphorics of the human scale material world simply breaks down. There is no mental representation for such entity, so quantum is just a quantity, or better, *the smallest increment* into which many physical properties are subdivided or the fundamental unit of *something measurable*. In other words, it is almost nothing but at the same time a 'difference that makes the difference' – which is Gregory Bateson's characterization of 'the elementary unit of information' [Bateson, 1973, p. 428].

7 Digital metaphysics

So, it is as close the pure difference as possible – or is it the pure difference itself? In case the infinitesimal close turns into the 'thing itself' the boundary between ontology and epistemology is erased and the way opened for 'digital (meta)physics': the epistemic abstraction of pure difference is ontologized into the ultimate constituent and consequently the 'old' notions of *matter and energy reduced into information*. Or as J.G. Chaitin, the developer of the 'algorithmic information theory', boldly puts it:

18 Prigogine and Stengers characterize the constellation as follows: '*Quantum mechanics thus presents a reversal of perspective relative to the classical style. It is no longer a question of looking for simplicity at the level of elementary behavior. Dynamic simplicity, as reflected by the possibility of a completely monadic representation, belongs, in fact, to the macroscopic world, to the world of our scale. Our physics is a science created by macroscopic beings, created with conceptual tools and instruments that belong to the macroscopic world. It is from that position, when we question the world of quanta, that we must choose what will allow us to express matters in terms of measure, reproducible, and communicable properties. We can no longer allow ourselves, as far as the physical world is concerned, the privileged point of view, which, when pushed to its limit, we once could have identified as that of God*'. [Prigogine and Stengers, 1982, pp. 135-155].

> Newtonian physics is now receding into the dark, distant intellectual past. It's not just that it has been superseded by quantum physics. No, it's much deeper than that. In our new interest in *complex* systems, the concepts of energy and matter take second place to the concepts of information and computation. And the continuum mathematics of Newtonian physics now takes second place to the combinatorial mathematics of complex systems [Chaitin, 2003].

Actually this implies a shift into metaphysics reminiscent of Leibniz's 'monadology' characterized by Prigogine and Stengers as follows:

> we know that for Leibniz the physics of aggregates of bodies affecting each other had only an imaginary character. It is a dream (but a coherent one) to attribute the variations of force at each point to external factors. That is to say, *in reality no longer from a physical point of view but from a metaphysical one, the world is composed of unextended substantial elements* – monads – each of which displays a predetermined internal law... [Prigogine and Stengers, 1982, p. 142, emphasis added].

No wonder the representatives of digital metaphysics respect Leibniz as the originator of their discipline. Chaitin even proposes ('not entirely seriously') the renaming of the *bit* into *leibniz* [Chaitin, 2003]. Anyway, formulated into a metaphysical scheme the *bit* really acquires the status of the *originary difference*, no different of the status of the *word* in the Judeo-Christian creation [St. Johns Gospel, see above]. And consequently it repeats the same unsolvable (supplementary) act/actor-configuration as well. This could be illustrated by quoting John A. Wheeler, another representative of digital (meta)physics, among others [Fredkin, Toffoli, Zeigler, Steinhart, etc.]:

> ...every particle, every field of force, even the space-time continuum itself – derives its function, its meaning, its very existence entirely – even if in some contexts indirectly – from *the apparatus-elicited answers to yes-or-no questions*, binary choices, bits [...] every item of the physical world has at bottom – at a very deep bottom in most instances – an immaterial source and explanation; that which we call reality arises in the last analysis from the posing of yes-no questions and the registering of equipment-evoked responses; in short, *that all things physical are information-theoretic in origin*... [Wheeler, 1990, p. 5, emphasis added].

To paraphrase St. John (1.1.): in the beginning there was a Bit (binary choice), and the Bit was with almighty Hacker, and the Bit was

Hacker, or simply the Supreme Computer (*Deus qua Machina*)[19]. Surely one may subscribe to such digital metaphysics without asking questions about the original agent. The Supreme Hacker can be bracketed and excluded just like God was pushed out of picture three centuries ago: the Creation was separated from the Creator and the epistemic object of nature or universe gained the status of an autonomous construction or system to be scrutinized.

However, it should be noted that digital metaphysics, represented by Wheeler (*et al.*) is a rather extreme venture which approximates in some instances mythical, mystic and religious ideas. Thus it is not a surprise to find some religious circles taking part in discussions around the topic. Likewise it should made clear that even though representatives of quantum computation are involved in these discussions many of the them reject such metaphysics. One of these is David Deutsch (from the Centre for Quantum Computation, University of Oxford) who replaced Wheeler's 'It From Bit' hypothesis (according to which the physical entities arise from underlying information content) with his version 'It from Qubit' (quantum+bit) in an article with the same title [Deutsch, 2003]. Thus he refuses to take the infinitesimal but fateful (perhaps even fatal) step into pure difference:

> nobody know how to derive *it* (the nature of the physical world) from *bit* (the idea that information plays a significant role at the foundations of physics), and I shall argue that this will never be possible. But we can do the next best thing: we can start from the qubit [Deutsch, 2003, p. 1],

which is 'the simplest type of quantum observable' that of 'a Boolean observable – defined as one with exactly two eigenvalues'. And he continues: 'This is the closest thing that quantum physics has to the classical programmer's idea of a Boolean variable' [*ibid.*, p. 3]. In Deutsch's scheme the qubit is still 'a type of *physical system*, rather than a *pure abstraction*' [*ibid.*, p. 4, emphasis added] – the latter being (infinitesimally) close to my notion of pure difference. So, it might

19 There are different ways to locate or replace God in the digital metaphysics. Gods place is given either to the Ultimate Programmer, the Ultimate Program (software) or the Ultimate Computer (hardware) running it.

just be that the fine line between the artifact and the nature is still there…

Bibliography

G. Bateson, *Steps to an Ecology of Mind*, Paladin/Granada, London 1973.

H. Blumenberg, *Der Prozess der theoretischen Neugierde*, Suhrkamp, Frankfurt am Main 1984.

F. Capra, *The Web of Life. A New Synthesis of Mind and Matter*, HarperCollins/ Flamingo, London 1997.

J. G. Chaitin, *Leibniz, Information, Math and Physics*, Kirchberg, 2003 <http://www.cs.auckland.ac.nz/CDMTCS/chaitin/kirchberg.html>.

J. Derrida, *Dissemination*, The University of Chicago Press, Chicago 1981.

D. Deutsch, 'It from Qubit', in J. Barrow *et al.* (eds.), *Science and Ultimate Reality*, Cambridge University Press, Cambridge 2003.

P. Dyer, *Revolutionizing the Sciences – European Knowledge and its Ambitions, 1500-1700*, Palgrave, Hampshire 2001.

I. Hacking, *Representing and Intervening*, Cambridge University Press, Cambridge 1983.

I. Hacking, *The Social Construction of What?*, Harvard University Press, Cambridge (Mass) 1999.

M. Heidegger, *Einführung in die Metaphysik*, Niemeyer, Tübingen 1953.

R. Kaplan, *The Nothing that Is – A Natural History of Zero*, Penguin Books, London 1999.

C. Lévi-Strauss, *The Savage Mind*, Oxford University Press, Oxford 1996 [1962].

N. Luhmann, 'Frauen, Männer und George Spencer Brown', in *Zeitschrift für Soziologie*, 17(1), 1988, pp. 47-71.

A. Meigs, 'Food as Cultural Construction', in *Food and Foodways*, 2, 1988, pp. 341-357.

F. Nietzsche, 'Jenseits von Gut und Böse', in K. Schlechta (ed.), *Friedrich Nietzsche – Werke in Sechs Bänden*, Carl Hanser, München 1980 [1886], pp. 563-759.

I. Prigogine and I. Stengers, 'Postface: Dynamics from Leibniz to Lucretius', in J. V. Harari and D. F. Bell (eds.), *Hermes: Literature, Science, Philosophy – by Michel Serres*, Johns Hopkins University Press, Baltimore 1982, pp. 135-155.

I. Prigogine and I. Stengers, *Order out of Chaos, Man's New Dialogue with Nature*, Collins, Glasgow 1985.

M. Sahlins, 'Other Times, Other Customs: The Anthropology of History', in *American Antropologist*, 85(3), 1983, pp. 517-544.

C. E. Shannon, 'A Mathematical Theory of Communication', The Bell System Technical Journal, 27, 1948, pp. 379-423, pp. 623-656.

J. Siikala, *Akatokamanava – Myth, History and Society in ther Southern Cook Islands*, The Polynesian Society, Auckland 1991.

P. Simons, *Parts. A Study in Ontology*, Clarendon Press, Oxford 1987.

G. Spencer Brown, *Laws of Form*, The Julian Press, New York 1972.

G. Steiner, *Grammars of Creation*, Faber & Faber, London 2002.

H. Tellenbach and B. Kimura, 'Some Meanings of the Concept "Nature" in Vernacular Languages and Their Correspondences in Japanese', in *International Philosophical Quarterly*, 19(2), 1979, pp. 177-185.

B. Timmermans, 'The Originarility of Descartes's Conception of Analysis as Discovery', in *Journal of the History of Ideas*, 60(3), 1999, pp. 433-447.

V. Valeri, *Kingship and Sacrifice – Ritual and Society in Ancient Hawaii*, The University of Chicago Press, Chicago 1985.

J. A. Wheeler, 'Information, Physics, Quantum: the search for links', in W. H. Zurek (ed.), *Complexity, Entropy, and the Physics of Information*, Addison-Wesley, Reading (Mass) 1990, pp. 3-29.

Pasi Falk, Helsinki Collegium for Advanced Studies, University of Helsinki

Karamjit S. Gill

The dance of the user in artificial cultural space

Abstract:

How can we know the dancer from the dance? *How is it that we can distinguish the one from the other, know the latter in the former without reducing it to or inferring from the particulars comprising it? These questions raise the issue of the relationship between actuality (e.g. music as comprising it?) These questions raise the issue of the relationship between the parts and the whole. 'How we can know the music from the music being played by the musician'? How can we know the music in the musician without reducing the music (actuality) to reality (e.g. aggregation of explicable notes of the music), or inferring from the particulars comprising it. On the similar lines, we can ask the question, how can we know the user from the contexts in which the user acts and interacts, or how can we model the user without reducing the dynamic inter-relationship of the diversity of user contexts to the discernable particulars of the context? The paper explores the interactions of the user in overlapping and interlinking cultural architectures. It will argue that the notion of the valorised cultural horizon enables us to extend the concept of 'interface' from the realm of the 'separation' between the reality and the actuality, to the symbiotic realm in which diverse cultural realities collaborate, by pooling together their praxes, experiences, knowledges and cultural resources, thereby enhancing both their objective capacities and tacit capabilities. In this perspective, the symbiosis represents the possibilities, choices and alternative paths available to human societies. This perspective of 'valorised reality' provides a methodological handle for bridging the gaps between* 'actuality' *and* 'reality' *mediated by the human-machine symbiotic interface. The interface is viewed here as the medium for supporting communication and dialogue between and within overlapping knowledge networks.*

Keywords:
actuality, artificial landscape, cultural architecture, holons, knowledge architecture, user interaction, valorisation

1 Knowing the user from the use context

The paper deals with one of the key issue of knowing the user form the use context. In this context we can ask the question how can we know the user from the contexts in which the user acts and interacts? or how can we model the user without reducing the dynamic interrelationship of the diversity of user contexts to the discernable particulars of the context? In comprehending the place and role of the user in various social, functional, and cultural contexts, we face the issue of understanding the dynamics of interactions between and among the individuals, groups and the community. In exploring the interactions of the user in various contextual, we recognise the need to understand the user interactions within user environments, including interaction at the intersections and boundaries of these environments. For example when a person moves from a functional environment (e.g. work) to a social environment (e.g. home), some of the user interactions may still be performed at a functional level while others take place in the social context. The same would be true of the user moving from the social to the functional space. We can say that the user interaction belongs to both to actuality of the situation (in which user acts) and reality of the situation in which we from the outside perceive the user to be acting. In this sense the user interaction can be described as an interplay between how we perceive reality and how we live in the actuality. Here we conceptualise the user interaction in two ways, either within the context of observed reality (objectified reality), or we can conceptualise it as a continuum between the observed reality and the actuality. In the first case we pursue a methodological approach, which accepts the separation between the objective space and the tacit space, while in the second case it recognises the symbiotic relationship between the actuality and reality in the sense of how life is (actuality) and how life should be (reality). The interface is viewed here as the medium for facilitating communication and dialogue between actuality and reality.

2 What then are the issues of user interaction?

By focusing on observed reality, we limit the user interaction to only one part of the user context, thereby excluding the other part which lies in the actuality space, the living context of user interaction. This implies that our conceptualisation of the user interaction space and therefore the design of user interface is bounded by the objectified reality. In other words thc interface design accepts separation between reality and actuality as the starting point of technology design.

This separation between reality and actuality is rooted into the scientific rationality, which subscribes to the separation between the objective and the tacit domains of user interaction. This leads to designing user interaction tools and user interfaces, which can only deal with first order interactions (based on observed data) and are too limited to deal with the second order interactions arising from the impact of the first order interactions, and third order interactions arising from the cumulative impact of first and second order interactions.

The user from this perspective is thus conceptualised as the functional actor in the utilitarian sense, therefore weakening and even excluding the user interactions within the actuality of the social and cultural contexts. In other words the focus of user interaction and interface design is conceptualised primarily in terms of functional roles such as client, patient, and taxpayer (minimum of abstraction according to Negrotti), and marginalises the tacit roles of the user as a social and cultural being, a member of the community and a citizen.

The implication of this separation of the reality from actuality is that the user is modelled only within a part of the interaction space, limiting it to see the user interactions within observed reality only. This conceptualisation of the user is rooted in the scientific rationality of the 'cause and effect', and leads to designing systems and tools in the 'one best way' methodological tradition. We can say in this perspective that this techno-centred model of the user can neither see the user nor hear the user within the holistic space reality-actuality. In other words we cannot know the user from the use context if we follow the techno-centric model of the user. To overcome the limitations of this techno-centric model of the user, we propose a human-centred model of the user, which is cultivated in the notion of the interdependence between the reality and the actuality, and thus forms the core of the notion of the 'symbiosis'. In the human-centred tradition, the symbiosis seeks an interdependent relationship between the social roles (tacit dimension) and the functional roles (objective) of the user, and thus between the actuality and reality. The notion of the symbiotic interface is perceived as the 'between' interaction between the reality (objective world) and the actuality (tacit, practice).

3 Conceptualising the user and user interface

From the discussion above, it can be argued that the model of reality (observed/explicated) can only be weakly applied to the interaction and interfacing domains in the world of the actuality (practice). This argument is based on the proposition that in comprehending reality, we first construct a model, which represents facts as we observe, and then use this model to design technology. However, in comprehending actuality, we experience actuality as it is practiced and experienced from within. The model of actuality thus represents the practice (tacit dimension). In other words the conceptual gap (between-ness) between reality (experienced) and actuality (experiencing), is further widened by models of their representations. We therefore need to cultivate a design culture, which overcomes the conceptual limit (due to 'between-ness') and cultural errors (due to the focus on the 'one best way') of the techno-centric model. This requires rethinking of the interfacing architecture for user interaction over the user interaction spaces. Basically the interface architecture can be defined in the following terms:

a. Interface between the machine (reality) and the machine (reality);
b. Interface between the human (actuality) and the machine (reality);
c. Interface between the human and the human

where reality is defined as an objectified/explicit representation of the situation and the 'actuality' as the experience in and of the situation. The interface then is 'in-between' interaction between situations.

The predominant view of interfacing tools tends to perceive the designer as the modeller of reality, and the user in terms of the user model as the representation of reality. In this sense, both the designer and the user are seen as the constituents of the same reality. The argument is that since reality can be objectified and represented as a technical model, so the designer and user can also be seen as integral components of the technical model. It is this convergence of the designer and the user in the technical realm which fails to recognise the 'between-ness' of reality and actuality, of the object and practice, and of the objective and the tacit. In this scenario we find that the interfacing (collaborating) technology design remains essentially concerned with the interface between the machine and the machine in the sense of objectified reality. The limitation of this technical determinism thus is that the methodology of tool design remains rooted in the separation of the objective from the tacit, and the separation of reality from actuality. It is this exclusion of the human situation (actuality) from technology design which leads to a growing concern about the place and role of the user and user communities in the debates on collaborating technologies or more properly on technologies for collaboration. For example, the information technologist may perceive the user in terms of research into human-computer interaction, the focus of the cognitive technologist may be on the cognitive models of mind, and AI scientist may regard the user in terms of knowledge mining. In a similar fashion, the philosopher may focus on the ethical dimension of the user, the sociologist on the societal dimension, the anthropologist on the cultural dimension, the engineer on the human-computer collaboration, the management scientist on knowledge management. At the same time an inter-disciplinary researcher and cross-disciplinary practitioner involved in the study of the impact of the convergence of digital technologies may see the user in terms of many social, cultural, functional and abstract roles depending upon the contexts of both the user and user interaction.

4 Actuality and the symbiotic interface

The argument on actuality draws upon Uchiyama's recent work on 'The Theory and Practice of Actuality' [Uchiyama, 2003] and the notion of symbiosis draws on the human-centred traditions [Cooley 1987, Gill 1996, Rosenbrock 1990]. In understanding the concept of actuality, Uchiyama notes that when we play and hear music being played, we not only hear what is being played but also use our past experiences of music and sounds as well as sounds we expect to create in future to hear the music as a 'whole'. In other words we hear music as a whole, in a sort of 'time', in which the 'past' and the future are always present (and absent) in the very 'now' of notes which are being played. It is this notion of the music as 'whole', which according to Uchiyama, resides 'between' real notes and us, in *actuality*. In clarifying the distinction between reality and actuality, Uchiyama draws upon the work of the eminent Japanese psychiatrist, Bin Kimura. Uchiyama notes that Kimura claimed that self exists in a duality of the real self and of the actual self. The 'real self' is a *neomatic* or 'experienced self' and the 'actual self' is *neotic* 'experiencing self'. In a healthy person, the real and the actual self progress in an interdependent way, drawing upon each other. However, in a schizophrenic, the actual self often becomes engulfed by the real self, leading to excessive self-reflection and disorder. It is this notion of the interdependence between the reality and the actuality, which is at the heart of the notion of the 'symbiosis'. In the human-centred tradition, the symbiosis seeks an interdependent relationship between the human (tacit dimension) and the machine (objective), between the actuality and reality in the Kimura sense. The notion of the symbiotic interface is perceived as the 'between' interaction, between the reality (objective world) and the actuality (tacit, practice).

Uchiyama draws a distinction between the way Western and Japanese participants comprehend situations. While the Western participant 'sees' the situation and relates to it as an objective observer, the Japanese participant 'hears' the situation, and relates to it by feeling to be 'in the situation'. In the first case, the interaction between the observer and the situation is through information, and in the second case the interaction is through language. In the human-centred tradition, we call the 'information'-oriented perspective of comprehension as the 'techno-centred' paradigm of technology design and the 'language'-oriented perspective as the 'human-centred' paradigm of technology design. In the techno-centric paradigm, the situation is modelled as an explicit representation of the observed reality, and this model of simulated reality is used to design interfacing technologies. Since the model in this case is the same as the reality in the factual (data) sense, there is no gap between the two realities, and hence there is not much scope of learning, either of learning by cognition or learning by doing. In the human-centred paradigm, the situation is modelled as a symbiosis between the reality (objective) and the actuality (tacit). It is this symbiosis which provides the 'between' framework for designing interfacing architectures. Drawing upon work on dialogical modelling and the definition of knowledge in terms of the objective, personal and experiential [S. P. Gill, 1995], the 'tacit' is seen here as the inter-relationship between the 'personal' (feeling/experiencing) and 'experiential' (collective experience/practice).

This articulation of the tacit provides a conceptual handle to articulate interdependent (symbiotic) relationships between the 'personal', the 'experiential' and the objective. It can be argued that part of the 'personal' knowledge can become part of the 'experiential' dimension over time during the process of participation in a group, and that part of the 'experiential' knowledge can become absorbed into the 'objective' dimension over time through the process of collaboration. Following the similar argument, it is proposed that part of the 'objective' knowledge can also be transferred to the 'experiential' domain, and part of the 'experiential' knowledge to the 'personal' domain. It is further proposed that this symbiotic idea of transference between 'personal' 'experiential' and the 'objective' provides an insightful framework for designing interfacing architectures for 'in-between' interactions between the cultural holons. It is this symbiotic notion of transference and 'between-ness', which provides an interdependent relationship between cognition and action, and thus the core concept of interfacing and collaboration. Central to the human-centred paradigm are the notions of the 'symbiosis' and 'purpose'. In summary, the essence of the argument here is that the notion of symbiosis provides a conceptual basis to design interfacing architectures for the 'between' interactions, and the notion of purpose acts as a catalyst and focus for designing collaborating technologies. Thus in seeking interdependence between the reality and actuality, between the objective and the tacit, the 'symbiosis' provides a conceptual handle to find coherence between diversities, ambiguities and uncertainties of the human situations modelled by cultural holons.

5 The actuality, reality and in-betweeness

To get further insight into the actuality-reality gap, we consider here how our comprehension of reality and our experience of actuality, leads to building interfaces which deal with the continuum or in-betweeness of user spaces of interaction. The argument is based on the proposition that in comprehending reality, we first construct an objective model, which represents facts as we observe them, and then we use this model to design interaction architectures. However, in comprehending actuality, we experience actuality as it is practiced and experienced [Uchiyama, 2003] from within the societal context. The model of actuality thus represents the practice (tacit dimension), which is situated in the social. In other words the conceptual gap (between-ness) between reality (experienced) and actuality (experiencing) is further widened by models of their representations. Thus interaction architectures based on the model of reality (techno-centric model) can only be narrowly applied to the domain of actuality in the sense that it can only interact with the objectified component of the domain, thereby being excluded from its interaction with actuality dimension. It is in this context that we can say that the interaction gap ('between-ness') between the actuality and the reality is not amenable to technical systems and tools, and it can only be amenable to human interaction. We can deduce form this argument that while technical systems can be designed for interaction within the domain of actuality, we need to design human-centred systems (symbiosis between humans and technology) for interaction within the domain of actuality. In other words our faith in the technical solutions for designing user interfaces is misplaced, and we need to focus on the design of interfaces, which seek symbiotic relationship between user spaces. We therefore need to cultivate a culture of design, which overcomes the conceptual limit of reality and seeks a continuum of interaction within, between and across user spaces of interaction.

The important issue here is to recognise that conceptualising societal interaction within a particular practice (actuality) can become limited by the techno-centric model of reality. While the actuality is rooted in the past experiences and is shaped by the present reality and future possibilities, the reality is defined by the observable facts and data as of 'now', the present. In an attempt to model the user interface, the designer is limited to objectifying the situation as 'observed', thereby excluding the many of the possibilities of the situation as 'being observed'. We thus see a further widening of the 'actuality gap' arising from the gaps between the actuality and reality of the both the user and the designer. For example in addition to the actuality gap, our interaction experiences within our social space, especially the tacit dimension, are 'transparent' to us most of the time, while they may remain hidden to users from other contexts, thereby limiting the inclusive conception of the user interaction from a cross-user perspective. Because of the actuality-reality gap, the rich interaction experiences resulting form social and cultural contexts of users, remain excluded form design of the techno-centric systems and technologies. It is this exclusion of the 'tacit' dimension of the actuality, which limits the design of creative and imaginative applications of experiential knowledge as a learning process, and thereby impoverishes the ability of a society to in shaping an holistic agenda for the knowledge society. The challenge therefore is to rethink about the appropriateness of the 'reality-centric' agenda as the dominant driving force of human development, and develop strategies and processes to mitigate its impact through societal concepts associated with actuality such as diversity, human purpose, participation, equality, social responsibility, ethics, and creativity. One way to move forward is to conceptualise knowledge in its many multifaceted roles, while recognising that the it is the interdependence between reality and actuality which facilitates the production, use, reproduction and creation of knowledge.

It is this notion of the interdependence between the reality and the actuality, which is at the heart of the notion of the 'symbiosis'. In the human-centred tradition, the symbiosis seeks an interdependent relationship between the human (tacit dimension) and the machine (objective), between the actuality (tacit, practice) and reality (objective). In the techno-centric paradigm, the situation is modelled as an explicit representation of the observed reality, and this model of simulated reality is used to design knowledge systems. Since the model of reality in this case is the same as the observed reality in the factual (data) sense, there is very little gap between the representations of two realities, and hence there is very little scope of learning, either of learning by cognition or learning by doing. In the human-centred paradigm, the situation is modelled as a symbiosis between the reality (objective) and the actuality (tacit). It is this symbiosis, which provides a conceptual tool for designing knowledge architectures. Drawing upon work on dialogical modelling and the definition of knowledge in terms of the objective, personal and experiential [S. P. Gill, 1995], the 'tacit' is seen here as the inter-relationship between the 'personal' (feeling/experiencing) and 'experiential' (collective experience/practice). This articulation of the tacit provides a conceptual handle to articulate interdependent (symbiotic) relationships between the 'personal', the 'experiential' and the objective. It can be argued that part of the 'personal' knowledge can become part of the 'experiential' dimension over time during the process of participation in a group, and that part of the 'experiential' knowledge can become absorbed into the 'objective' dimension over time through the process of collaboration. Following the similar argument, it is proposed that part of the 'objective' knowledge can also be transferred to the 'experiential' domain, and part of the 'experiential' knowledge to the 'personal' domain. It is further proposed that this symbiotic idea of transference between 'personal' 'experiential' and the 'objective' forms the core catalyst for designing communication architectures for knowledge networking and cross-cultural collaboration. In summary, the essence of the argument here is that the notion of symbiosis provides a conceptual basis to design knowledge architectures for the cross-cultural interaction, and the notion of purpose acts as a catalyst for designing systems and technologies for cross-cultural collaboration. Thus in seeking interdependence between the reality and actuality, between the objective and the tacit, the 'symbiosis' provides a methodological tool to find coherence between diversities, ambiguities and uncertainties of the human condition.

6 Rethinking the knowledge architecture

The networked vision of the information society recognises the need for building new models of collaboration at local and global levels for dealing with changing nature of information and knowledge flows within and across cultures, regions and societies [Castells, 1966]. At the same time, there is also a recognition that the overlapping and intersecting nature of information and knowledge networks requires a rethinking about the development architectures which deal with multidimensional nature of communication and interaction within and between these networks, at the local, global and local-global levels. In this architecture we see the tacit dimension situated in the local and the objective (explicit) situated in the global. Knowledge networking is thus seen in terms of seeking symbiotic relations between diversity of local knowledge architectures and the global architecture. Since the tacit represents the diversity and complexity of local cultural architectures, we need to comprehend the diversity of cultural architecture in order for us to comprehend the communication relationships between and across the local and global architectures. For example, even if the diagnostic knowledge of a particular domain (e.g. particular crop or particular disease) may be shared among the scientific community at the global level, its application and implication of its use at the local level is dependent upon diversity of social, cultural, religious and political contexts. In other words the interfacing between the global knowledge and local knowledges is much more than building technological architectures for communication, and it is to do with building communication architectures, which map onto interactions between cultural architectures at local and global levels.

To illustrate the nature and complexity of this interaction, we consider exemplar models of cultural holons (Fig. 1), ranging from the one-one interaction in a Linear Holon; dialogical interaction in the 'Trinity' Holon, and to interlocution in the 'Dragon' Holon. The idea of the cultural holon is that just as the model of reality is used to represent the environment as it is observed and experienced, so a cultural holon can be used as a model to represent the architecture of a culture as it is observed in interaction. Here cultural holon is seen as one of the many possible ways of modelling and representing cultural architectures. The holon here represents the interconnectedness of relationships between and among the units and whole of the cultural architecture – a model of the cultural uni-verse [Koestler, 1989]. For example, the 'linear' holon represents the linear architecture, which can be said to exist in highly, individualised cultural architectures (e.g. technocentric), but of course linear architectures can either be found to exist or created in almost all cultures. The 'trinity' holon representing interaction between three nodes allows for mediation by the third node in case of disagreement or conflict between any two nodes within cultural holon. It can be regarded as the minimal communication model of a democratic dialogue within the holon. For example, to reach a democratic agreement or consensus within the Trinity Holon as a whole, an agreement or a consensus must be reached at each levels of interaction, both at the hierarchical and horizontal levels of interaction between the nodes of the holon. The 'panchtantra' holon representing interaction between five nodes (interconnectedness of two trinity holons), represents the next level of complexity of interaction than the trinity holon. In this case, we begin to see interaction between nodes at both the vertical and horizontal levels, thereby representing a much more robust model of democratic dialogue. The next two cultural holons, the 'smaurai' and the 'dragon' represent further levels complexity of cultural architectures. We see that as the complexity of cultural architecture increases, we need to build more and more complex interaction architectures in order to deal with the diversity of interaction.

To illustrate the complexity and problematics of interaction between the holonic architectures [Mathews, 1995], let us consider the interaction between the Linear Holon and the Trinity Holon. One possible way is to deconstruct the Trinity Holon in three linear holons, thus creating three levels of interaction between the Linear Holon and three linear components of the Trinity Holon. This is possible only if we regard each of the three linear components of the Trinity Holon as independent linear holons. This implies that interaction between the nodes of each linear component is independent of the structural links and contextual relationships with other nodes and other components of the Trinity Holon. This is to say that the only way to construct interaction between the Linear Holon and the Trinity Holon (and for that matter any other more complex holon) is reduce the complex structure of the Trinity Holon to the level of the Linear Holon. If we follow this reductionist logic to its own logical conclusion, then what we are implying is that interaction between different cultures can take place only if we either reduce the more complex cultural architectures to map onto the technological architectures of interaction, or we reduce both the cultural architectures to a common denominator amenable to technological architecture. This reductionist logic so inherent in the techno-centric paradigm reduces culture to a technical artefact, thereby reducing cross-cultural interaction to a technical interface. It is therefore hardly surprising that the dominant presence of the techno-centric model of development rooted in the rationality of the 'one best way' and convergence paradigm, leads to a belief in technological solutions of complex societal problems. If we continue to pursue this techno-centric path, knowledge systems and development may not be an exception to this logic.

Given the dominant place of the information science model of interaction promoting direct interaction at the individual levels, it is important to note that the sustainability of interaction within these linear architectures can only be sustained if both the nodes are always actively engaged in interaction and there is no breakdown in communication between the nodes. Although the linear architecture may be regarded as the most cost effective architecture at a direct interact ional level and most explicit from the individual perspective, it remains in danger of instability and breakdown without continuously being serviced at all costs. Since this architecture has no structural mechanism for third party mediation or intervention in case of uncertainty or conflict, it therefore lacks any room for diversity and democratic coherence necessary for participatory development. We already are beginning to see social and cultural vulnerability inherent in the linear interactive network communication structures (e.g. internet) promoting direct communication between individuals without social filtering or cultural mediating structures. There also lies an ethical dilemma of 'governance' of technological architectures and their operations as they impact societies and cultures.

If we regard these holons as representative models of diverse cultural architectures, then the challenge is:

a. How to design knowledge systems that facilitate interaction and collaboration within each cultural holon; between cultural holons, recognising that interaction within and between more complex cultural holons takes place at a multiple levels of hierarchy, both horizontal and vertical;
b. How to design knowledge architectures (e.g. cognitive, social, cultural) that facilitate collaboration within and between cultural holons related to the above challenge of 'actionable knowledge' is the conceptual/methodological challenge of developing cross-cultural models of collaboration and knowledge architectures for human development.

7 Valorisation of communication spaces for collaboration

In the overlapping and interlinking networked world, there is a need to create a dynamics of local-global communication spaces in which:

a. The user (self) interacts with the other user (other), and in the process they create a shared interaction space (symbiotic space), while remaining located in their own local spaces (Fig.2). It is assumed hat both the local spaces belong to same cultural space;
b. The user (self) belonging to interaction space A (Reality A) and the 'other' belonging to the 'other' interaction space (Reality B), create a new third interaction space (Reality C), 'space of the artificial' in order to valorise their differences and reach a common understanding. In this scenario, the common space of the artificial (Fig. 3) is evolved for seeking a common purpose, while retaining own cultural identifies rooted in local interaction spaces. This model of the space of the artificial could be adopted to create future common communication spaces (culture of the artificial space) for interaction and collaboration of cultural holons.

In the symbiotic space (Fig. 2), the objective knowledge of both the local spaces can be harmonised to form part of the common interaction space. As a result of the ongoing interaction, parts of the tacit knowledge of both the users become part of the shared knowledge, thereby becoming part of the evolving symbiotic space. At the same time part of the tacit knowledge of the users becomes part of the shared experience and ultimately part of the objectified knowledge over time, thereby expanding the objective knowledge space. Similarly the process of sharing of knowledge experiences leads to gaining new experiences in both the interaction spaces, thereby expanding the experiential (tacit) knowledge of each of the user spaces. In this scenario of a dynamic symbiotic space, both the experiential knowledge and the objective knowledge spaces expand and thereby enrich interaction.

In the space of the artificial (Fig.3), we envision a scenario in which users from two different interaction (cultural) spaces (Reality A and Reality B) come together by creating a new 'third' reality C, the space of the artificial, or culture of the artificial' [Negrotti, 1999]. This can be seen as finding a symbiosis of two realities (cultures), providing a common communication space for common purpose and mutual understanding. This 'third' culture, an artificially created cultural space, initially created for bringing two cultures together in a harmonious relationship, slowly takes the shape of a new 'third' cultural space, where conflicts can be shared and resolved, while respecting the diverse identities and divergences of two cultures, thereby enriching both the artificial (may also be called 'virtual') cultural space and the cultural spaces (realities) of the users. In this scenario, cultural realities can be seen to be valorised through shared communication. The 'third' reality (space of the artificial) may also be called as the 'valorised reality'. The notion of valorisation emphasises the making 'the best' of the differences during the process of reaching an agreement, a common purpose, or finding a common ground. Valorisation thus enables the processes of seeking cohesion of diversities (e.g. social, economic, cultural) while recognising and respecting differences. For example, in collaborating situations, participants from diverse cultures may create a new different cultural space (cultural of the artificial space) in which they engage in dialogue and interlocution in finding a common ground of their differences, while at the same retaining their unresolved difference (identities) to themselves (i.e. located in the own cultural spaces). This process of 'valorisation' allows for continuous enhancing of cross-cultural collaborating (or engagement) spaces while at the same enriching the own cultural horizons of participants.

This valorised cultural horizon enables us to extend the concept of 'interface' form the realm of the 'separation' between the reality and the actuality, to the symbiotic realm in which diverse cultural realities collaborate, by pooling together their praxes, experiences, knowledges and cultural resources, thereby enhancing both their objective capacities and tacit capabilities. In this perspective, the symbiosis represents the possibilities, choices and alternative paths available to human societies. This perspective of 'valorised reality' respects plurality of cultures, and seeks coherence through empathy and harmony. 'Culture of the artificial' in the sense of the transfiguration of reality [Negrotti, 1999] provide a methodological handle for shaping and building of artificial landscapes by bridging the gaps between '*actuality*' and '*reality*', mediated by the human-machine symbiotic interface. The interface is viewed here as the medium for supporting communication and dialogue between and within overlapping human and virtual networks.

8 Designing an operational architecture for cross-cultural interaction

This valorised cultural horizon enables us to extend the concept of 'interface' form the realm of the 'separation' between the reality and the actuality, to the symbiotic realm in which diverse cultural realities collaborate, by pooling together their praxes, experiences, knowledges and cultural resources, thereby enhancing both their objective capacities and tacit capabilities. In this perspective, the symbiosis represents the possibilities, choices and alternative paths available to human societies. This perspective of 'valorised reality' provides a methodological handle for bridging the gaps between '*actuality*' and '*reality*' [Uchiyama, 2003], mediated by the human-machine symbiotic interface. The interface is viewed here as the medium for supporting communication and dialogue between and within overlapping knowledge networks.

The issue here is how to design an operational model of interaction for the cultural architecture represented by cultural holons, which enables interaction i) between users within a given cultural holon, ii) between users (or user groups) from different cultural holons to interact in a shared virtual cultural space while sustaining their distinct cultural identities, iii) between and among intersecting and overlapping cultural holons in which users from different holons create shared communication spaces dependent upon value structures of their cultural holons. One of the structures, which is seen to be relevant to get an insight into the interaction at both horizontal and vertical levels, within and across cultural holons (Fig.4), is holonic architecture [Mathews, 1995].

Holons represent the interconnectedness of relationships between and among human systems, between the unit and whole – a model of the uni-verse. In the holonic interface model (Fig. 3), each holon (A, B, C & D) represents a closed interaction space; the interaction between holonic spaces (say A & B) takes place through representative nodes (users). The interaction between two holonic spaces can be seen in terms of seeking a symbiosis between these two interaction spaces.

In this scenario, there is a possibility of creating a symbiotic interaction space between two holons (as in Fig. 2) or the evolution of the 'space of the artificial' (as in Fig. 3) between two holons (their user representatives).

We can extend this process to create two holonic interaction spaces, say holonic symbiotic space X between holons A & B, and holonic symbiotic space Y between holons C & D. Both these holonic interaction spaces X and Y can then be valorised to create the next level of holonic interaction space, say Z. This process of symbiotic interaction spaces provides a model for the creation of dynamic interaction spaces at various levels of interaction between holons and between holonic interaction spaces, thereby providing a model for designing holonic interfaces between holons and networks of holons [Gill, 2002].

9 Concluding remarks

Conceptualising the user within the domain of observed reality limits the conceptualisation of the user interaction and use interface to functional (objectified) part of the user interaction space. In other words this conceptualisation is only limited to dealing with the first order interactions, bounded by the observed reality.

This focus on the separation of the reality from the actuality creates a conceptual gap between the objective and tacit dimensions of user interactions. This conceptual gap leads to designing user interaction systems and interfacing tools, which remain impoverished by their design to deal with the second order user interactions arising from the impact of first order interactions.

It is only by conceptualising the user within the holistic vision of user interaction space covering the actuality-reality domain that we begin to know the user from user interactions covering the continuum of the user interaction space defined by the multiplicity of the social, functional and cultural roles of the user. This conceptualisation, covering the in-betweeness, provides a conceptual framework for designing symbiotic interfaces for users within the same interaction context, as well as, for users belonging to different interaction contexts.

This conceptualisation gives rise here to two scenarios:

> How to design interface which deals with in-betweeness of two different user interaction contexts, and
> how to design interface which deals with in-betweeness of two different user interaction contexts at multiple levels of interaction, both horizontally and vertically.

In the case a), the notion of the 'culture of the artificial' provides a conceptual basis to create an third virtual interaction space which provides a medium for evolving commonality of interactions (common purpose) of users belonging to two separate interaction spaces, while allowing the users to deal with their own user specific interactions within their own use contexts. We term this conceptualisation as the 'valorisation' of interactions.

In the case b) we propose a network architecture of user interactions at various levels of interaction, 'linear' holonic structure dealing with interaction between two users, 'trinity' holon dealing with interactions between three different user interaction spaces, the 'panchtantra' holon dealing with five different spaces, 'samurai' holon dealing with seven different spaces, and 'dragon' holon dealing with nine different interaction spaces.

Finally we propose a holonic architecture as framework for designing symbiotic interfaces, which deal with interactions between and across user interaction spaces, both at horizontal and vertical levels of interfacing.

This human-centred approach brings together the conceptual notions of actuality, the essence of the culture of the artificial, the notions of the symbiosis and the holon, and provides both a conceptual architecture for designing and operationalising user interfaces at various multiple levels of user interactions within and across multiple levels of user interaction spaces.

Bibliography

AI & Society: Journal of human-centred systems and machine intelligence, Springer-Verlag.

M. Castells, *The rise of the networked society*, Blackwells Publishers Ltd. Oxford 1996.

M. J. Cooley, *Architect or Bee?*, Hogarth Press, London 1987.

K. S. Gill, *Human machine symbiosis*, Springer, London 1996.

K. S. Gill, 'The user in the emerging space of the digital provide', *Proceedings of the Conference on 'The User Models in the Advanced Society'*, Venice University, IUAV (Venice University Institute of Architecture), 18-19 November 2002.

S. P. Gill, *Dialogue and tacit knowledge for knowledge transfer*, PhD Dissertation, University of Cambridge, Cambridge 1995.

A. Koestler, *The ghost in the machine*, Arkana, London 1989.

J. Mathews, 'Holonic organisational architectures', in *Human Systems Management*, Vol. 15, 1995.

M. Negrotti, *The Theory of the artificial*, Intellect Books, Exeter 1999.

H. H. Rosenbrock, *Machines with a Purpose*, OUP, Oxford 1990.

K. Uchiyama, *The theory and practice of actuality*, Daito Bunka University Press, Tokyo 2003.

Karamjit S. Gill, IRIHCS, Brighton and IMES-LCA, Urbino

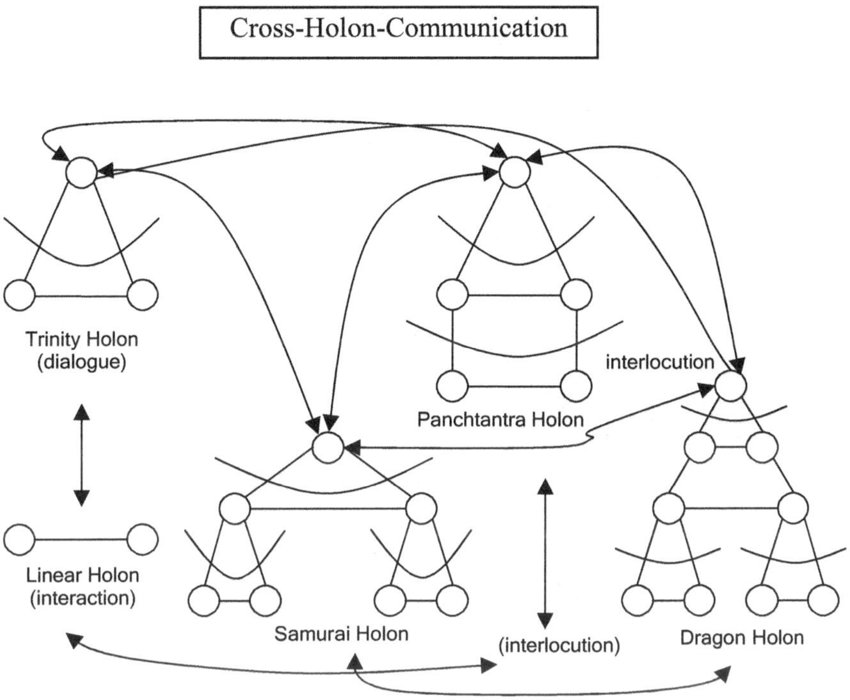

Fig. 1. Holonic Architecture of cross-cultural interaction illustrating interaction and communication modes between cultural holons, each cultural holon as a simplified representation of the observed reality of interaction within a cultural structure of communication

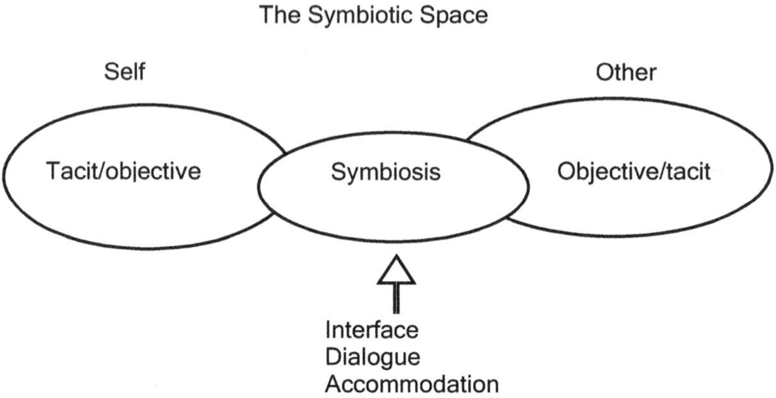

Fig. 2. The evolving symbiotic space through the symbiosis between the local (tacit) and global (objective), seeking a common space of interaction within a context

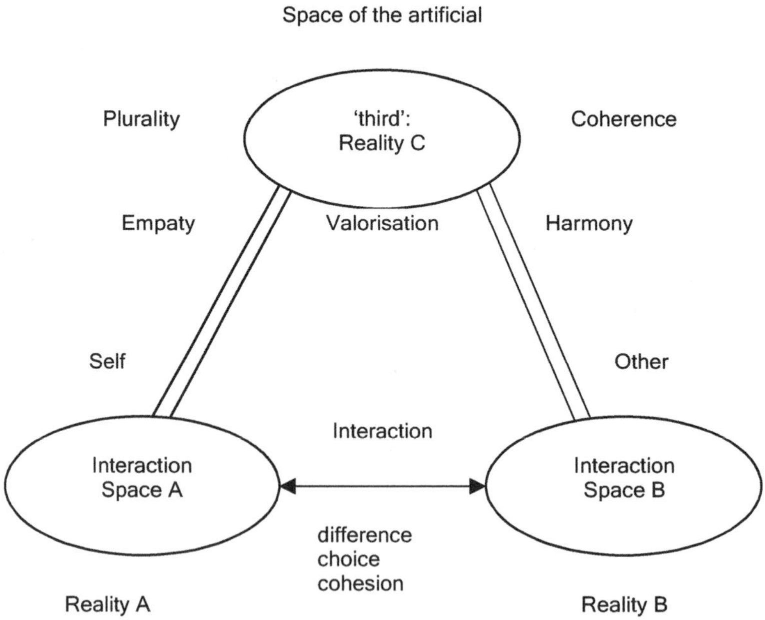

Fig. 3. The creation of the space of the artificial, the third reality (common ground), for cross-cultural understanding and common purpose, created by valorisation of two different realities

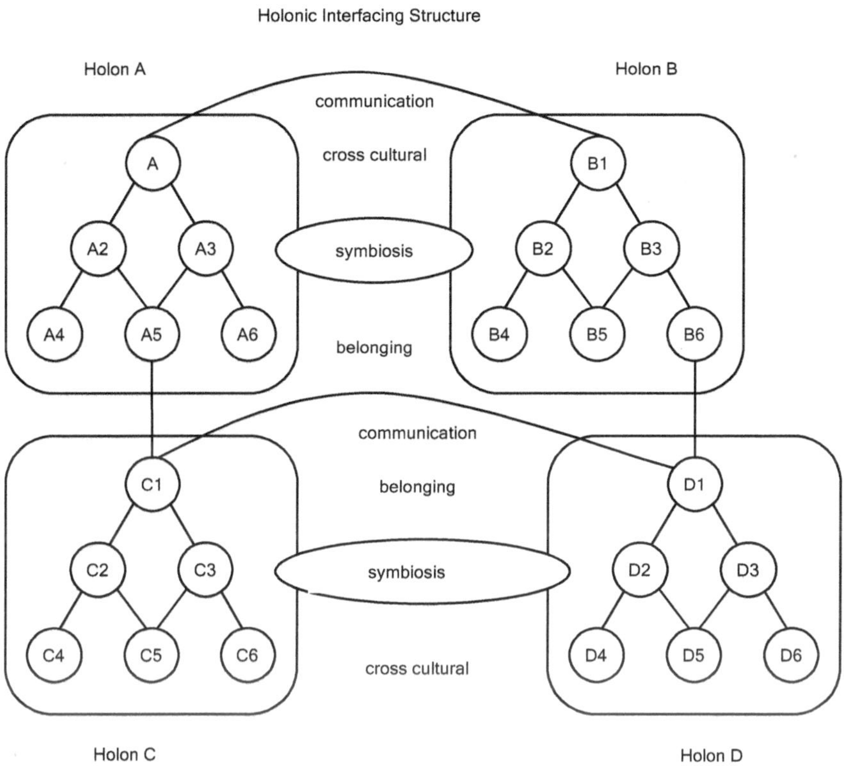

Fig. 4. Holonic interfacing architecture, illustrating the vertical interaction between holons through their representative nodes, and horizontal interaction through a symbiotic interface between the two interacting holons

Stefano A. E. Leoni

Le diverse et artificiose machine … to make music, or: Different and artful machines … to make music

Abstract:

*The title of this paper refers to Agostino Ramelli's wellknown text (*Le diverse et artificiose machine*); Captain Agostino Ramelli's machines were 'diverse et artificiose' (different and artful), the machines I'm showing produce squared artifices, 'squared machine o artifact (effective or theoretical) to make music'. Being a musicologist I don't discuss the* vexata quaestio *on the definition of 'Artificia'l but I want to speak here about technological devices, about 'machines-ready-to-make-music' mostly during the Baroque, a period highlighting a trend to reification of art, of the work of art, with a resulting artificialization of the product.*

After a wide survey on some early automatic music instruments and composition machines, the paper presents several organs by Kircher and his Arca Musarithmica, *a device for automatic composition and some other curiosities like the German flute-clocks of 18th-19th centuries for which even Haydn and Beethoven composed music. What kind of music? Interesting but not emotional for a 'User', or without any real user.*

Keywords:
music, machine, device, water organ, flute-cloks, arca musarithmica, artifices

The title of this paper refers to Agostino Ramelli's wellknown text (*Le diverse et artificiose machine*) published in Paris in 1588; Captain Agostino Ramelli's machines were 'diverse et artificiose' (different and artful), the machines I'm showing you produce squared artifices. I should like to speak in short about a particular side that I could define 'squared machine o artifact (effective or theoretical) to make music'. Being a musicologist I don't discuss the *vexata quaestio* on the definition of 'Artificial'; we agree we can draw a distinction between technology of the artificial and conventional machines (conventional technology): as Negrotti states 'unlike the latter, the former implicitly or explicitly, aims to reproduce something existing in nature'. Likewise on the possibility of producing a naturoid going over the imitation of natural model. Neither I'm referring to theories defining artificial every material modification produced on cultural bases. Nor I mean to speak here on the communicative property of music (that would generate artificial realities 'where transfiguration is present then') or to deny it (there is no artificial in music, but the original as it is); all these topics were sharply and cleverly discussed by Massimo Negrotti, such as for the 'musical observation' too.

I want to speak here about technological devices, about 'machines-ready-to-make-music' mostly during a historical-aesthetic period, the Baroque, highlighting a trend to reification of art, of the work of art, with a resulting artificialization of the product. I want to speak about Athanasius Kircher, odd clever scholar, half the inventor of an immense virtual, literary Wunderkammer (a Cabinet of Curiosities), half the organizer of a proto-encyclopaedia: his *Musurgia Universalis* (Rome, 1650).

Kircher's automatic machines produce music as a sounding background, they are machines that don't hear anything, that have to work knowing what they are doing, whatever the use and the user of their production are. Functional music takes no part in artistic remarks: functional music fulfil its duties, apart from the fact that 'someone shaves, has a wash, changes his clothes, sleeps, reads, thinks, etc.; musical machine is indifferent, it doesn't know why, nor for whom, it works: it works and that's enough'.

Just as a wide survey, I remind here only some early automatic music instruments and composition machines, such as the Hydraulis, the Aeolian harp, the medieval Isorythmic motet, the 16th century Soggetto Cavato motivic technique.

The *Hydraulis* was invented by Ktesibios in the second century B. C. Ktesibios, was fascinated by pneumatics and wrote an early treatise on the use of hydraulic systems for powering mechanical devices. His most famous invention, the Hydraulis, used water to regulate the air pressure inside an organ. A small cistern called the pnigeus was turned upside down and placed inside a barrel of water. A set of pumps forced air into the pnigeus, forming an air reservoir, and that air was channeled up into the organ's action.

Greek *Aeolian harp*. This may be considered the first automatic instrument. It was named for Aeolus, the Greek god of the wind. The instrument had two bridges over which the strings passed. The instrument was placed in a window where air current would pass, and the strings were activated by the wind current.

Isorhythmic motets were developed in the late Middle Ages (14th-15th centuries). These songs made use of patterns of rhythms and pitches to define the composition. Duration and melody patterns, the *talea* and the *color* respectively, had not the identical number of elements. Music was developed by the different permutations of pitch and rhythmic values. So if there were 5 durations and 7 pitches, the pitches were lined up with the durations. Whatever pitches were 'left-over', got moved to the first duration values. The composer would permute through all pitches and durations before the original pattern would begin again.

Soggetto cavato, is a technique of mapping letters of the alphabet into pitches. This technique was used Josquin's Mass (late 15th century) based on the name of Hercules, the Duke of Ferrara. One application of soggetto cavato would involve be to take the vowels in Hercules as follows: e=re=D; u=ut=C (in the solfege system of do, re, mi, fa, etc., ut was the original do syllable); e=re=D. This pattern of vowel-mapping could continue for first and last names, as well as towns and cities.

Not only: during the Renaissance the first mechanically driven organs and water organs were built. In 1624 the English philosopher Francis Bacon wrote about a scientific utopia in the *New Atlantis*. Here he stated: «we have sound-houses, where we practice and demonstrate all sounds, and their generation. We have harmonies which you have not, of quarter-sounds, and less slides of sounds ». In 1650 Athanasius Kircher described in his *Musurgia Universalis*, a mechanical device that composed music. He used number and arithmetic-number relationships to represent scale, rhythm, and tempo relations, called the *Arca Musarithmica*.

More: in 1761 Abbe Delaborde constructed in Paris a Clavecin Electrique; Wolfgang Amadeus Mozart composed the *Musikalisches*

Würfelspiel (Musical Dice Game) in 1787. This composition was a series of precomposed measures arranged in random eight-bar phrases to build the composition. Each throw of a pair of dice represented an individual measure, so after eight throws the first phrase was determined. And so on to the Panharmonicon by Maelzel, the inventor of the metronome in the early 1800s.

Let's now look after a surprising jesuit, the father Athanasius Kircher. In his book, *Musurgia Universalis*, we can find a thorough description of the automatic water-powered organs of the Italian renaissance as well as his automatic device to compose music, the *Arca Musarithmica*: practical, handly automata and theoretical automata. The first ones are some organs[1].

Fig. 1. Scheme of organ from Kircher's *MusurgiaUniversalis*

1 Some texts come from the interesting web site <http://web.telia.com/~u13101111/kircher/index.html> were you can find also pictures and musical examples.

The main part of the instrument has the form of a grotto, in which Pan the satyr plays his flute, accompanied by the cuckoo's call and the rooster's crow. In addition, there is a smaller organ with an effigy of a nymph, known in the Greek mythology as Echo; so the music of this organ has an eco effect.

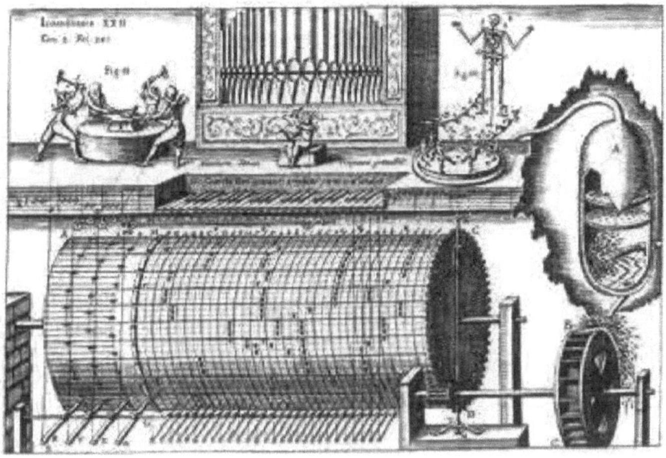

Fig. 2. Scheme of organ from Kircher's *MusurgiaUniversalis*

According to Kircher, this large organ has three ranks of pipes, and plays a four-part hymn accompanied by moving figures. The three blacksmiths are a tribute to Pythagoras, who is said to have discovered the harmonic intervals when listening to the sound of hammers in the forge. In addition, the organ's own music is enhanced by a four-part choir singing a Latin text invoking the blacksmiths' mythical patron Vulcanus and praising the glory of Work.

The automatic organ of the Roman Quirinale was built in 1647 by Italian organ builder Matteo Marione, following the designs of Athanasius Kircher. The present organ replaced an older instrument, built in 1596 by Luca Blasi. Since then, it has been restored several times and is still playable. The Quirinale organ is powered by water.

Water is use to turn the paddle-wheel which drives the drum with the melody, which moves the keys of the organ, and to provide the wind which enters the chest and is used to make the pipes sound.

Fig. 3. Automatic organ of Quirinale

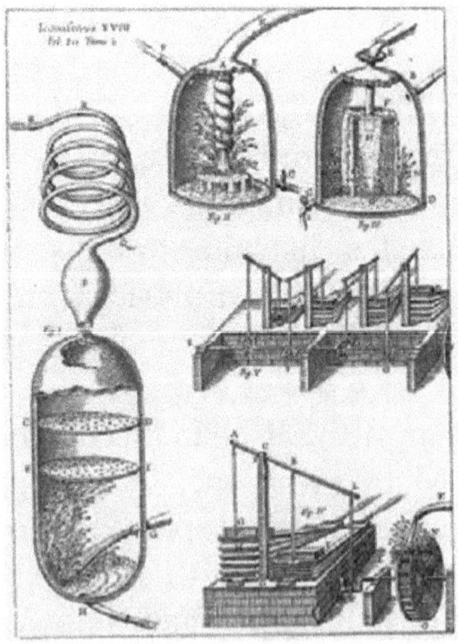

Fig. 4. Mechanism of water organ, from Kircher's *Musurgia Universalis*

Using a technique which was also employed in the iron-foundries of Italy as early as the the sixteenth century, water is made to enter a long vertical tube, usually with a funnel-shaped opening. In doing so, the water forms a gyre, and sucks air into the tube as well, forming an air-water mixture. At the end of its descent through the tube, the water is separated from the mixture in a special container ('Aeolian chamber'), and may be used to drive a paddle-wheel, as in the case of this organ. The air is allowed to exit via another tube which is connected to the windchest. No bellows are involved – the pressure is regulated by controlling the flow of water into the system.

Fig. 5. Arca musarithmica, Pepys Library of Magdalene College Cambridge

In his book, Athanasius Kircher gives us detailed instructions for pinning the barrel, which he called the cylindrus phonotacticus. He also gives us numerous examples of music appropriate for automatic organs, one of which is a four-part *Ricercata* by Johann Caspar Kerll (1627-1693).

And now a theoretical automatic-artificial instrument, by Kircher again. The *arca musarythmica* is a device by which a non-musician could compose a piece of four-part music using prearranged musical fragments inscribed in wands arranged in columns inside the box. Each type of wand corresponded to a particular metrical unit e.g. 4, 5, or 6 syllables, and on each wand there were examples of florid counterpoint on one side and more simple note-against-note settings on the other. Once the phrase to be set had been analyzed into its fundamental syllabic units, each of these could be set to an example taken from a wand of the appropriate type. There are many arcas still extant, including one in the Pepys Library of Magdalene College Cambridge.

Kircher's machines play and imitate (squared artificial machines) and compose; the latter is the Arca musarythmica as we saw: all the rest is 'combinatory'. The rest is the attempt to disclose the secret of music, to formalize not only its basic rules, but also its sense producing activity.

I think of one of Captain Gulliver's Travels, the one in Laputa, at the Academy of Lagado.

> The first Professor I saw was in a very large Room, with forty Pupils about him. After Salutation, observing me to look earnestly upon a Frame, which took up the greatest part of both the Length and Breadth of the Room, he said perhaps I might wonder to see him employed in a Project for improving speculative Knowledge by practical and mechanical Operations. But the World would soon be sensible of its Usefulness, and he flattered himself that a more noble exalted Thought never sprung in any other Man's Head. Every one knew how laborious the usual Method is of attaining to Arts and Sciences; whereas by his Contrivance, the most ignorant Person at a reasonable Charge, and with a little bodily Labour, may write Books in Philosophy, Poetry, Politicks, Law, Mathematicks and Theology, without the least Assistance from Genius or Study. He then led me to the Frame, about the Sides whereof all his Pupils stood in Ranks. It was twenty Foot Square, placed in the middle of the Room. The Superficies was composed of several bits of Wood, about the bigness of a Dye, but some larger than others. They were all linked together by slender Wires. These bits of Wood were covered on every Square with Paper pasted on them, and on these Papers were written all the Words of their Language, in their several Moods, Tenses, and Declensions, but without any Order. The Professor then desired me to observe, for he was going to set his Engine at Work. The Pupils at his Command took each of them hold of an Iron Handle, whereof there were fourty fixed round the Edges of the Frame, and giving them a sudden turn, the whole Disposition of the Words was entirely changed. He then commanded six and thirty of the Lads to read the several Lines softly as they appeared upon the Frame; and where they found three or four Words together that might make part of a Sentence, they dictated to the four remaining Boys who were Scribes. This Work was repeated three or four

Times, and at every turn the Engine was so contrived that the Words shifted into new Places, as the Square bits of Wood moved upside down[2].

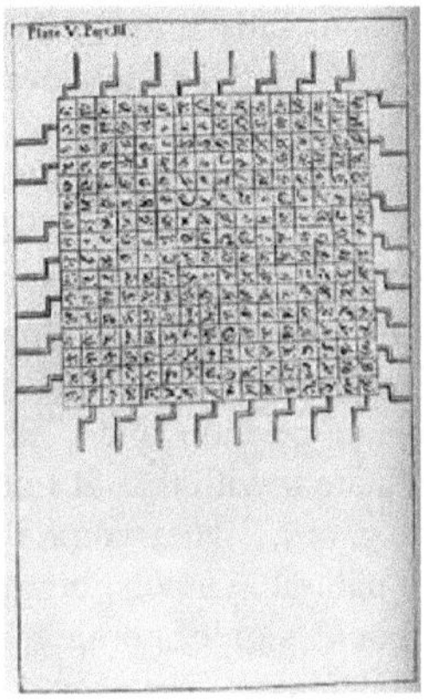

Fig. 6. Lagado Academy machine as Swift pictured it

Let's look for one of Kircher's *Tabulae mirificae*:

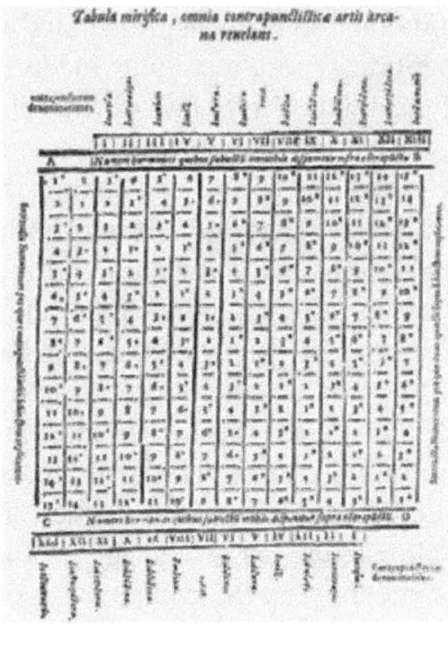

Fig. 7. Tabula mirifica from Kircher's Musurgia Universalis

2 J. Swift, *Gulliver's Travels*, Penguin Books, London 1994 [1726], pp. 200-201.

Something like Jorge Luis Borges' *The Library of Babel*:

> [...] Second: *The orthographical symbols are twenty-five in number*. This finding made it possible, three hundred years ago, to formulate a general theory of the Library and solve satisfactorily the problem which no conjecture had deciphered: the formless and chaotic nature of almost all the books. One which my father saw in a hexagon on circuit fifteen ninety-four was made up of the letters MCV, perversely repeated from the first line to the last. Another (very much consulted in this area) is a mere labyrinth of letters, but the next-to-last page says *Oh time thy pyramids*. This much is already known: for every sensible line of straightforward statement, there are leagues of senseless cacophonies, verbal jumbles and incoherences. (I know of an uncouth region whose librarians repudiate the vain and superstitious custom of finding a meaning in books and equate it with that of finding a meaning in dreams or in the chaotic lines of one's palm ... They admit that the inventors of this writing imitated the twenty-five natural symbols, but maintain that this application is accidental and that the books signify nothing in themselves. This dictum, we shall see, is not entirely fallacious).
>
> [...] Five hundred years ago, the chief of an upper hexagon came upon a book as confusing as the others, but which had nearly two pages of homogeneous lines. He showed his find to a wandering decoder who told him the lines were written in Portuguese; others said they were Yiddish. Within a century, the language was established: a Samoyedic Lithuanian dialect of Guarani, with classical Arabian inflections. The content was also deciphered: some notions of combinative analysis, illustrated with examples of variations with unlimited repetition. These examples made it possible for a librarian of genius to discover the fundamental law of the Library. This thinker observed that all the books, no matter how diverse they might be, are made up of the same elements: the space, the period, the comma, the twenty-two letters of the alphabet. He also alleged a fact which travelers have confirmed: *In the vast Library there are no two identical books*. From these two incontrovertible premises he deduced that the Library is total and that its shelves register all the possible combinations of the twenty-odd orthographical symbols (a number which, though extremely vast, is not infinite): Everything: the minutely detailed history of the future, the archangels' autobiographies, the faithful catalogues of the Library, thousands and thousands of false catalogues, the demonstration of the fallacy of those catalogues, the demonstration of the fallacy of the true catalogue, the Gnostic gospel of Basilides, the commentary on that gospel, the commentary on the commentary on that gospel, the true story of your death, the translation of every book in all languages, the interpolations of every book in all books.
>
> [...] The impious maintain that nonsense is normal in the Library and that the reasonable (and even humble and pure coherence) is an almost miraculous

exception. They speak (I know) of the 'feverish Library whose chance volumes are constantly in danger of changing into others and affirm, negate and confuse everything like a delirious divinity'. These words, which not only denounce the disorder but exemplify it as well, notoriously prove their authors' abominable taste and desperate ignorance. In truth, the Library includes all verbal structures, all variations permitted by the twenty-five orthographical symbols, but not a single example of absolute nonsense. It is useless to observe that the best volume of the many hexagons under my administration is entitled *The Combed Thunderclap* and another *The Plaster Cramp* and another *Axaxaxas mlö*. These phrases, at first glance incoherent, can no doubt be justified in a cryptographical or allegorical manner; such a justification is verbal and, *ex hypothesi*, already figures in the Library. I cannot combine some characters

dhcmrlchtdj

which the divine Library has not foreseen and which in one of its secret tongues do not contain a terrible meaning. No one can articulate a syllable which is not filled with tenderness and fear, which is not, in one of these languages, the powerful name of a god. To speak is to fall into tautology. This wordy and useless epistle already exists in one of the thirty volumes of the five shelves of one of the innumerable hexagons – and its refutation as well.

[...] Perhaps my old age and fearfulness deceive me, but I suspect that the human species – the unique species – is about to be extinguished, but the Library will endure: illuminated, solitary, infinite, perfectly motionless, equipped with precious volumes, useless, incorruptible, secret[3].

A little overview on other musical automata, like the Pomeranian Chest, built in 1617 by Achilles Langenbucher of Augsburg, this amazing piece of artwork was presented to Duke Philip of Pomerania. It contained an automatic organ with twenty-one wooden pipes playing four long pieces of music, A similar chest, containing an automatic virginal, was presented to Swedish king Gustavus Adolphus and is now a centerpiece in the collections of Upsala University.

Or like the more familiar Black Forest flute clocks. The clockmaking industry in the German Schwarzwald, or Black Forest, originates from the late 17th century. Apart from the well-known cuckoo clock (and its lesser-known relative, the trumpeter clock) the production embraced all known types of musical clock, including xylophone, carillon, flute and harp. There are a lot of musical examples for

3 J. L. Borges, *La Biblioteca de Babel*, Buenos Aires 1941.

'Schwarzwälder Flötenuhren', like a popular book by Herbert Jüttemann.

Also the great Austrian composer Joseph Haydn (1732-1809) is known to have written music for the Flötenuhr, or flute clock.There are 32 pieces composed by, or attributed to Haydn, who was then court musician to Prince Eszterházy of Eisenstadt, Hungary.

Ludwig van Beethoven (1770-1827) too wrote several pieces for automatic organs; not only the famous 'Wellington Victory', originally composed for a great automatic organ. Beethoven provided the environmental music for the Waxworks of Count Müller-Deym in Wien. He wrote five pieces for flute-clock.

We can hear all these pieces, even by Haydn or by Beethoven, and we can define them 'interesting', but …

But, to conclude, let me recall Athanasius Kircher, my main point of reference:

> The Musurgia mirifica shows its good working as a assembly set of prearranged musical sequences that finds its power point in being inspired to combinatory techniques: even having a limited number of musarithmi at his disposal, the composer is able to carry out ever-varied chord sequences connections. It's a way to act that intrinsically avoids the repetition of already-used sequences. In my opinion the musurgia mirifica is a technique having revolutionary potentials: nevertheless the general structure of the project hasn't the organization that would have allowed to achieve entirely the target. In fact we can state that in these pages from the VIII book of the Musurgia Universalis are reported the surviving vestiges of an unsuccessful attempt: perhaps a constructive instrument with sophisticated possibilities would have flourish; it was reduced by its deviser himself to the rank of a parlour game for the sake of simplification and 'audience' or customer requirements or for a lacking consciousness in his action[4].

And what's about the User? I would like to 'play' a little on words: there's no User, or he's a gambler, a player (dice gambling, laputian machines, arcae musurythmicae, library of Babel, musical slot machines, etc etc); a player (a mere performer: i.e. to play the piano) or,

4 C. Chierotti, 'La *musurgia mirifica* di Athanasius Kircher: la composizione musicale alla portata di tutti nell'età barocca' in *Nuova Rivista Musicale Italiana*, n. 3, 1994.

more, an angel (had an angel, at the beginning of the World – Kircher says – started combining the numbers he wouldn't have finished yet today – in 1650). An *anghelos*, i.e. a messenger, who inevitably suggests us the world of communication, and, why not, the idea of entropy such it's used in theory of information (Shannon): what's the amount of entropy we can measure in these musical bits? I presume it's high, such the amount of lacking information (in the aesthetic sense) is.

Maybe this is the reason why the User is at large. And maybe this is the reason why everything else is music but 'automatic one': Brahms' opus 114 *Trio*, as well as *Ruby Tuesday* by the Rolling Stones.

Bibliography

J. L. Borges, *La Biblioteca de Babel*, 1941, in *Ficciones (1935-1944)*, Sur, Buenos Aires 1944.

C. Chierotti, 'La *musurgia mirifica* di Athanasius Kircher: la composizione musicale alla portata di tutti nell'età barocca', in *Nuova Rivista Musicale Italiana*, n. 3, Rome 1994.

A. Kircher, *Musurgia Universalis*, Georg Olms, Hildesheim-New York 1970 [Rome 1650].

M. Negrotti, *L'osservazione musicale. L'artificiale fra soggetto e oggetto*, F. Angeli, Milano 1996.

M. Negrotti, *Naturoids. On the Nature of the Artificial*, World Scientific, Singapore 2002.

J. Swift, *Gulliver's Travels*, Penguin Books, London 1994 [1726].

Web site: <http://web.telia.com/~u13101111/kircher/index.html>.

Stefano A. E. Leoni, Conservatory of Music of Torino and IMES-LCA, University of Urbino

Renata Livraghi

The user of educational services

Abstract:

The user is a person expressing a demand for services. In the past there was no prevailing production of services, and the figure of the user was introduced to make a distinction from the consumer and the entrepreneur who expressed a demand for material goods and capitals. The users, on the contrary, demand services and immaterial capitals (education, health and so on). Today the prevailing production is that of immaterial goods and the processes of accumulation of immaterial capitals are acquiring greater importance than those for accumulation of material capitals. The figure of the user thus becomes more complex, assuming different roles and objectives in relation to the theoretical model to which it refers. Here we shall analyse the role of the user of educational services in three theoretical models. The user will therefore be considered as an 'agent' in the model of human capital, as an 'interested party' (stakeholder) with a strategic role in the managerial model that seeks maximum quality, and finally we will have the user of educational services in the capabilities approach *whom we indicate as a 'person'. It follows that the user is utilizable in given contexts and in theoretical models that, however, require clear identification as, otherwise, the user, in an analysis of* knowledge economy, *can no longer be seen as an operator able to explain how the economic system evolves.*

Keywords:
human capital, stakeholders, capabilities approach, quality, community of practice

1 Introduction

The user is one who applies for services [Porrello, 2003]. We are proposing here to analyse the demand for educational services expressed by a user we could define as an 'agent', by a user we could define as a 'stakeholder', by a user who may be a 'person'.

The need to define the user in different ways arises in relation to the theoretical model of reference, and so varies according to the ultimate purpose which led to the demand for education.

In the human capital model the user is an 'agent'; in managerial models, where the aim is maximum quality, the user is a 'stakeholder'; in the 'capabilities approach' the user is a 'person'.

The user, in the human capital model, applies for educational services in order to acquire that new knowledge and professional expertise needed for increasing their productivity. The usefulness of the educational services required by the user must retain their validity over medium to long-term periods of time, and will therefore involve an investment to improve the user's level of human capital. This in turn leads to increased productivity and therefore to higher wage. When viewed as an 'agent', the user applies for educational services in order to qualify for a higher income in the future [Johnes, 2000].

In managerial models, where the aim is maximum quality, the user takes an active part in planning, providing and evaluating the educational project. In these models several 'stakeholders' may be identified in the educational process so that there are different users with different needs requiring educational services provided by a single process. Each user requires these services for particular needs. The users in an educational process are: students, teachers, the world of labour. All participate actively in the process while having different aims to pursue, though with a single purpose in mind, namely that of improving the quality of the services offered [Livraghi, 2002]. But what do we mean by a better quality of educational services? How can different needs be reconciled in pursuing a common aim, the results of which will subsequently be subjected to assessment?

In the 'capabilities approach' the user is a person who applies for educational services to acquire greater efficiency in different spheres of interest where active participation and understanding are needed. The user applies for educational services to better his professional level and be able to enjoy family and community life. In other words the 'person' requires educational services in order to achieve greater individual and collective wellbeing[1]. In this model educational services are not only requested to increase production for exchange through the mechanism of the market; the demand for these services also exists for improving that person's health enabling him to act in such a way as to raise the quality of an individual's life and that of a community [Livraghi, 2004]. The demand for educational services can therefore only exist in economic systems in which *lifelong learning for all* is considered as one of the basic *capabilities* for growth of human development within the community.

2 Users, investments and rationality

The demand for educational services in the three models is a demand for investment goods. Users seek education in order to acquire the efficiency, ability to work and live that bring about a growth of capital.

In the human capital model educational services are provided to further its growth as a resource for increasing production of goods and services exchanged in the markets; educational services are sought in order to accumulate human capital and therefore enjoy a better income at both individual and community levels.

In the managerial model educational services are requested to increase the human capital of the various stakeholders and to accumulate organizational and social capital. The purpose of demanding educational services is to accumulate human capital, organizational capital, social capital so that a higher income may be achieved at both individual and collective levels, to improve the quality of the educational product, to promote cooperation among the various parties concerned and thus to benefit social relations. In this model the demand for educational services does not generate human capital only, it also

1 *Wellbeing* here means the possibility of choosing to live a well-balanced and happy life in full awareness and responsibility. *Wellbeing* represents a wider concept than that generally applied to the term *welfare* which implies the availability of means for leading a respectable life at a decent level, but also opportunities and the capacity to act freely in the pursuit of worthwhile objectives.

generates a process of accumulation of organizational and social capital [Livraghi, 2003].

In the 'capabilities approach' model educational services are sought to increase the human capital of the individuals in a community as well as to increase their capacity for functioning and for empowerment. The demand for educational services is therefore an expression of the positive freedom of individuals. In this model the demand for education not only leads to increases in human capital and in social capital, it also benefits personal capability and therefore promotes the personal and cultural capital of the individual.

The demand for educational services therefore leads to an investment that increases immaterial capital: human capital, organizational capital, social capital, and the personal and cultural capital of the members of a community.

Attention of the classical economists was mainly devoted to the accumulation of material capital and it was not until the 1960s that immaterial capital, considered exclusively as human capital, was introduced into their models of economic analysis. They made studies of the direct effects produced by education while ignoring the indirect effects promoting accumulation of the other components of immaterial capital. Managerial models and the 'capabilities approach' made possible a realization of the interrelations existing between immaterial capital in its various forms and choices made by the users of educational services, in contexts of reality rather than of abstractions as required by models of human capital.

The word 'agent' is applied to the user in the model of human capital who is considered as an abstract individual that makes choices and takes decisions in relation to an explicit process of calculation comparing costs and benefits as if for making an investment in stocks and shares. In this model the user first evaluates the direct and indirect costs to be borne for investing in human capital with reference to the expected returns that should be greater than costs. In actual fact, the user of the human capital model never manages to obtain the complete information that the model requires. The choice has to be made, in a situation of asymmetrical information, between the likelihood of being able to carry forward the educational process without incurring educational 'debts' and that of finding employment adequate to the course undertaken. Another important limitation pertaining to the model of human capital lies in the mode of calculating the returns that an investment in education will provide. These returns vary over time according to the productive activity in which the user is involved, to the organizational environment in which such activity is operating, as well as to individual effort. Actual returns can therefore be very different from those estimated before deciding whether or not to apply for education. It follows that the user in fact is not an 'agent' but is a person to decides to make an investment where only asymmetrical information is available and where there is no certainty as that results will come up to expectations.

Managerial models transform users from passive to active subjects, to mitigate the deficiencies of the model of human capital. In models of this type we find users of different kinds who cooperate to secure high quality education in order to limit the risks of the investment in human capital and therefore obtain better returns. The users are known as 'stakeholders' in the education process; each one has different needs to satisfy and different aims to achieve. In the model of human capital there is only one user who applies for educational services, and who accepts the education offered on the assumption that it will provide exactly those skills required by the labour market.

In actual reality where each worker operates in a condition of asymmetrical information, the stakeholders should take joint action in the educational process so as to clarify the different needs of each one and, if possible, work out an educational project common to all. Some points of conflict may arise among stakeholders; the needs as stated by students may diverge from the requirements expressed by the other two parties interested in the educational process, namely by teachers and by the labour market. A single educational process may include, among other things, teachers voicing demands for different educational services: for example there may be students lacking operative experience and who, at the end of the educational process, want to find work in line with the knowledge they have accumulated, but also worker students who aim to improve their professional position and obtain a higher wage following participation in the educational process if the results achieved are recognised as satisfactory. Compared with the traditional type of student, worker students face greater obstacles when participating in the educational process; this because they have to gain recognition of the knowledge already acquired in their professionalizing working experience and therefore obtained in ways other than those provided by institutional types of teaching. Their main request is that the educational process shall be organized differently from that offered to full-time students, that the subjects taught and methods used shall be suited to their needs. A state of conflict may therefore arise between worker students and the other stakeholders (ordinary students, teachers, labour market). Worker students want personalised training courses able to satisfy particular and specific requirements, but it is unlikely that the other stakeholders concerned in the process can agree to provide these facilities. One difficulty is caused by the schools themselves where organizational models tend to be somewhat rigid, but also by teachers who are used to passive attitudes on the part of students and who are unwilling to cooperate on equal terms with the other stakeholders in the educational process. Worker students, on the other hand, ask their teachers to create an environment where knowledge of a new kind can be provided; being already working adults they desire to take an active part in the educational process that will enable them to make a synthesis between theoretical knowledge and working experience. They therefore ask that the teacher-role in the educational process shall change: no longer purveyors of knowledge but creators of knowledge, interacting on equal terms with their students. The need therefore arises for creating communities of learning consisting of full-time students, worker students and teachers in which each user takes an active part. Technology can give a positive reply to the needs expressed by worker students, able as it is to overcome the limitations posed by space and time.

The labour market can make more information available to stakeholders by indicating, when the education program is being planned, what capabilities are most in demand and, during the manufacturing process, by cooperating with teachers and students over planning the training courses and carrying them out. Interaction between teachers/universities and the labour market helps to create new knowledge by encouraging research of various kinds, while employment of young people with a good level of education favours innovation in processes of production and in the products themselves.

New technologies and creation of communities of practices and of learning can therefore settle or at least mitigate possible conflict among the various users of managerial models; but this can only come about by continuous improvement in the professional capacity of teachers, by their willingness to accept and operate changes, and by their ability to cooperate with the different stakeholders in the educational process. In these models the continuous processes of self-assessment and of external assessment are essential as the education market is an asymmetrical market; special mechanisms are therefore required to check on quality and on the use made of available resources. If this happens the educational process creates new knowledge, improves the organizational structures of schools and of the firms that employ the students who complete their courses satisfactorily, establishes communities of practices, favours the creation of networks. Put briefly, the managerial models can favour the accumulation of immaterial capital and lessen the risks inherent in investments made in human capital. These models, however, necessitate continuous adaptation to change in people and in organizational structures, as well as an ability to learn that must certainly continue after the educational process has been completed. Lifelong learning should be one of the capabilities of the economic system for all those concerned in it.

The 'capabilities approach' is something recent. It emerges from the research work carried on by A. Sen and aims to bring out the direct and indirect benefits of education and of the other essential components of human development. Its analysis introduces the concept of capability that represents freedom to reach and obtain the combination of functionings that represent the lifestyles it is possible to achieve. Education and the chance to continue learning throughout life are therefore a 'capability'; the freedom a person actually enjoys for widening his or her possibilities of action, for doing those things believed to be worthwhile [Sen, 2000]. Continuous education is therefore a 'capability' of great importance to a person, alongside others such as, for example, the possibility of living a long and healthy life, of benefiting from an income sufficient for a respectable existence within one's own social environment; all this because it widens the extent of a person's overall freedom and is of importance for improving his chances of choosing and achieving the lifestyle that he would really like to have. The level of development of a society is in fact judged according to the substantial freedoms enjoyed by people and their liberty of action. If education increases a person's overall freedom and active participation in his own world, in those economic systems where people can take part in processes of continuous education we also find an increase in the personal and cultural capital of each single person.

Lack of 'capabilities' deprives people of the chance to accumulate the various forms of immaterial capital which constitute their sources of income and wellbeing.

In the 'capabilities approach' the user of educational services applies for them in order to act and acquire empowerment in his own community. The acquisition of empowerment means possessing elements of immaterial capital that become readily tangible because expressed in the professional skills achieved, in participation in associations, in communities of practice, as well as in other intangible parts of immaterial capital such as awareness of, and responsibility for one's own actions, in self-esteem, in a capacity for reflection and criticism, in the sense and purpose of life. A person possessing these personal characteristics can therefore recognise and overcome obstacles and be able to seize the opportunities that present themselves in the various stages of human life.

In the 'capabilities approach' the user of educational services is called a 'person' because the purpose of his demand is not only to acquire professional skills, as happens in the theory of human capital, and not only to demand a better level of education as happens in the managerial models of education. This user of educational services desires to widen his capabilities, or the different combinations of functionings open to him. In other words, in the 'capabilities approach' the user asks for education in order to achieve his own ends with the means available. This theory highlights the interconnections among the various types of immaterial capital and the capabilities of an economic system, making clear the behaviour of the various users and related results.

Both these aspects deserve separate attention even if often correlated. The capacity for action focuses greater attention on the person and what that person does. A person's capacity for action is not in fact usually independent of his state of wellbeing. In conventional models of economic analysis, on the other hand, a distinction is not made between these two aspects because economic development always coincides with the degree of wellbeing of a community.

In the 'capabilities approach' the demand for educational services enables us to realise the effects of education and learning both in individuals and at a social level.

3 Conclusions

In the past, the term user was adopted to distinguish the user from the consumer and from the investor who demanded material goods; the user, on the other hand, demanded services, namely immaterial goods difficult to estimate. Production in industrialised countries today consists mainly of immaterial goods, and the processes of accumulating immaterial capital are gaining in importance compared with the accumulation of material capital, formerly decisive for economic growth. Production of immaterial goods requires productive factors that differ from those used even in the recent past, a flexible organizational structure able to discover and enhance individual knowledge, to facilitate creation and spread of new knowledge and of social capital.

In those economies mainly based on knowledge or in those characterised by market failure or by asymmetrical information [Campiglio, 2002] it is therefore no longer necessary to distinguish the user from the consumer or from the investor since, in contexts such as those, there is a demand for personalized services of high quality requiring constant interaction with manufacturers, lifelong learning and acquisition of empowerment by the stakeholders concerned in a manufacturing process or in realization of a project. In these cases 'users' remain as before but are differently termed in relation to their particular theoretical model. The user can therefore be an 'agent' in one case, a 'stakeholder' in others, or a 'person' where they are recipients of both direct and indirect effects of the demand expressed, or where the causes that prevent need from becoming a demand are analysed. In this connection the demand for educational services provides an example of special interest and significance.

Bibliography

L. Campiglio, *Tredici idée per ragionare di economia*, Il Mulino, Bologna 2002.

G. Johnes, *Economia dell'istruzione*, Il Mulino, Bologna 2000.

R. Livraghi, 'La coerenza della riforma universitaria con i problemi posti dall'economia della conoscenza', in L. Frey and E. Ghignosi (eds.), *L'importanza della riforma universitaria in corso in Italia*, Quaderni di Economia del Lavoro, n. 74, Franco Angeli, Milano 2002, pp. 33-64.

R. Livraghi, 'Economia della conoscenza e capitale sociale', in L. Frey, R. Livraghi, and G. Pappadà (eds.), *Lavoro a elevato livello qualitativo e sistemi economici/sociali fondati sulla conoscenza*, Quaderni di Economia del Lavoro, nn. 76-77, Franco Angeli, Milano 2003, pp. 97-163.

R. Livraghi, *Sistemi economici e sviluppo umano*, report presented at the Seminar organized by the Istituto Rezzara, Recoaro Terme, 10-12 September 2004, now being published.

A. Porrello (eds.), *Modelli dell'utente nelle società avanzate*, Ed. IUAV, Venezia 2003.

A. Sen, *Lo sviluppo è libertà. Perché non c'è crescita senza democrazia*, Mondadori Editore, Milano, 2000.

Renata Livraghi, Dip. di Studi Politici e Sociali, Università di Parma

Giuseppe O. Longo

Narrative as a source of the artificial

Whence come, then, the pleasure or the compulsion of men to narrate stories?
Peter Bichsel

Abstract:

Narration is a constant activity of humans, by which they transform the given world into artificial, mental worlds that are simpler to live in. This construction is based on changing memories and dynamical plans, hence the artificial worlds built by narration are always changing. The arborescent structure of narration is more suited than the linear structure of the scientific discourse to represent the complex and contingent, dynamical nature of reality as perceived by human beings. Western science has always tried to discover the eternal and universal laws of nature, neglecting singular events, and history has gradually followed the same path. On the other hand, technology and politics are interested in singular artifacts and everyday events. Narration, intricate and arborescent as it is, is the best tool to reflect the intricate and arborescent process by which novelty, information, life and culture develop. Narration, technology and politics are questionable, corrigible, modifiable. They all strive to create new, artificial (mental or material) worlds.

Keywords:
artificial, complexity, contingency, dynamics, history, narration, memory, politics, process, science, simulation, technology

1 Narration and the construction of artificial worlds

We, as human beings, have always narrated stories to other people and to ourselves. This never-ending narrative, be it confined in ourselves or projected outwards to meet other humans, has two effects: first of all, it constructs a simplified image of the world, which permits us to survive its immense complication. There are several and diverse narrative techniques that we use to introduce order in the natural chaos: art, narration proper, science, technology. Their aim, basically, is to reconstruct the world, more precisely, to replace the 'given' world by an *artificial world*, simpler and more suited to us.

Secondly, through the narration process we try to construct a coherent and stable image of our 'self'. This construction is always doomed to failure, since the self is changing and ambiguous, multiple and fleeting. Still, this identification work is performed over and over again, since it is necessary. Every one needs to offer him- or herself and the others a unitary image, that is summarized by the pronoun 'I' and is the actor of our memories and the protagonist of our plans. Of course, the image of the world and the image of the self are strictly intertwined and are correlated to the aims that we pursue at each particular moment.

There is another reason for us to tell stories and to listen to stories: we possess only one life and we wish we had many. To overcome this unbeatable limit, or at least to have that illusion, we immerse ourselves into the alternative lives created by the narrative. Although we know, in some deep part of our consciousness, that they are fictional, we are eager to experience them as true, albeit ephemeral, lives and we want to inhabit those worlds that we do not possess. And the narrated, artificial life cancels the real one, at least for some time.

Memories and plans are the hub around which the narrations of the self revolve and such interaction is dynamical and reciprocal. Actually, on one hand the image we have of ourselves is modified when we recall memories and build plans, on the other hand memories and plans are modified by the self being formed. The necessity of coherence that the dynamical self exhibits implies a continuous modifications of our plans and of our memories as well. In this way, memories diverge more and more from their original source, and become almost independent of it.

2 The dynamical mutability of the narrative artificial

This complex interplay of memories and narrations is put into motion not only by memory, but also by oblivion. In fact, the transformation and sublimation of memories is the consequence of a complex interaction of memory and oblivion. Oblivion is a singular mechanism, a sort of beneficent selective filter that preserves and enhances positive memories and weakens or cancels negative ones. Hence, oblivion is not total and indiscriminate, rather selective and value-sensitive, emotion-sensitive, deeply rooted in our body and in our past. Oblivion clears the memory store, eliminating the wastes and accommodating the new experiences that in turn will become memories. Actually, oblivion operates a reshuffling of the various levels of observation and description, and a re-focusing of their importance. The artificial world and the artificial life built by narration are always changing.

Plans too undergo a continuous updating and rehandling through simulation, an activity that is typical of human beings as much as narration. Simulation is to plans more or less as narration is to memories. The simulation possesses a great economic and survival value for humans. Before we take a concrete action, usually we simulate it using our imagination or instruments that prolong or enhance it, like mathematical tools or computers. We can thus analyse the possible consequences of that action and decide whether to take it, modify it or give up. At the higher levels of intelligence and consciuosness, we perform mental tasks that refine the quality of any plan before we implement it. Narration and simulation concern methods and practices that belong to the world of information, not to the world of matter and energy. In the world of information, which is at the crossing point of many disciplines, such concepts as meaning, structure, form, relationship, and order hold sway. It is a world of description and explanation distinct form the world of collisions and forces studied by physicists: it is the world of communication, i. e. the world of living beings, in particular of human beings.

3 Simulation as source of the artificial

Simulation is a practice of artificiality. It can either have a real model (an 'objective' narrative or account of a fact in the world) or not have such a model (fictitious narrative). In the latter case the term 'artificial' may seem less appropriate, but it is not totally inappropriate because all narratives are made of components possessing elements of artificiality, i. e. imitation and reduction of real traces.

The boundary between narration and simulation is fuzzy: many of the stories we narrate are 'invented', i. e. are a sort of explicit simulation although they may use the past tense. But also the narratives that we define 'objective', i. e. reproducing facts that happened, are always partly modified by the filters of memory and oblivion and by the partiality of the viewpoint, and contain interpolations and extrapolations, i. e. simulated components inserted into a stuff that in principle is more 'real'. A fictional story perhaps exhibits what Peter Bichsel has called the 'irresponsibility' of the narration: we deal not with the truth, rather with the *possibility of the truth*. On the other hand we never invent anything, because 'literature only stems from literature, where there are no pioneers but only imitators that reflect' [Bichsel, 1989]. In other words, here the 'real' prototype is very very far away.

4 The discovery of uncertainty and complexity

The way we acquire our knowledge is changing. Western culture has inherited from the Greeks a predilection for abstract rationality and for theoretical rigour and has always considered the intelligence of mind, that builds the theorems of mathematics or the structures of metaphysics, as superior to the intelligence of the body. Our science is a product of this attitude.

The dream of the West, starting from Plato down to Einstein through Descartes and Leibniz, is to give a purely mental explanation of the world. But this dream has not come true. Recently our mind has been connected more and more intimately with more and more sophisticated technological research instruments, especially with computers, and the way we acquire knowledge has changed. The image we have of the world has changed, too: we discover a lot of uncertainty, complexity and disorder in it. Until the half of the 20th century (and even much later in the popular conception) the universe was thought of as possessing an extreme underlying simplicity, that sooner or later would be unveiled and summarized in one formula. Today the picture has changed: we are realizing that the regular order of reality assumed by physics, an order that we do not know yet but will certainly know in a not too far future, the existence of such a simple and clear order is a wishful thinking. Hence we have to resort to the narration of stories, a multiplicity of narrations reflecting an irreducible complexity and a multiplicity of viewpoints that reveal themselves as complementary, concurrent, antagonist, contradictory… A complex world demands complex stories.

To the multiplicity of viewpoints that complexity forces us to take into account correspond the plurality of history, the contingency of evolution, the importance of singular events, of bifurcations and of 'catastrophes', the many paths that all systems, including the universe, can take at any instant. Actually, *contingency* is an ineluctable ingredient of all phenomena. As a consequence the traditional logic of explanation, that only accepts what happened, considering it as necessary, and neglects what did not happen, considering it as impossible, should be replaced by the logic of contingency and of historical possibility, that explains what happened by comparing it with the various alternatives that could have occurred instead. To understand a phenomenon means to follow its genesis as opposed to the alternatives that contingency has eliminated, driving it on the path of irreversible existence, and to observe its surfacing from the waves of possibility, the only survivor of a thick bundle of admissible ramifications that have been pruned by the relentless shears of history.

5 Narration is indispensable

That is why the *narration of stories* is so important: stories are the only way to sense the lost possibilities and the contingency that transforms one and only one of them into an irreversible necessity, opening the way to further contingecies and thus conditioning, although feebly, the future. Stories are a mirror of History, because they have the same arborescent structure: in every narrative there are branching points, and the irreversibility appears only after the choice made by the narrator at such points. These choices, that open and close alternative worlds in the mind of the listener, are almost never obligatory, as we all know. Only in the development between one choice and the next, between a branching point and the next, a sort of determinism seems to hold sway in the story. This explains why it is only by a narration that we can understand and explain a phenomenon in its unfolding: a narrative has the form of what it narrates. The cold and unperturbed accounts of today science in its systematic stage lack all contingencies and all vivid germinative process, and are imprisoned in the rigid armour of logic reasoning. Storytelling is not a simple and rough activity preceding the refined and aseptic scientific account, rather it is a basic way of representing the world, which reflects the impossibility for the subject to represent reality from an hypothetical external, absolute viewpoint. There is no such neutral viewpoint, and the world can only be approached through a multiplicity of partial descriptions. We do not even know whether such descriptions can ever be unified into a single all-encompassing picture.

6 Narration, technology, politics

From the Greeks we have inherited an inclination for the heavenly perfection of 'ideas' together with a distaste for everyday earthly matters. This might explain the disgust that many intellectuals feel for politics, that is the realm of everyday life, of collective actions, of negotiations and details. Following the Greek, the Cartesian and then the idealistic philosophy, science has always strived to eliminate contingency, irregularity, plurality, and concreteness. Philosophy has tried to impose a totalitarian rule and to introduce a radically homogeneous way of thinking into politics. When it comes to understanding politics, philosophy is inadequate. Politics needs multiple stories, irreducible to one another, whereas the foundation of philosophy demands the elimination of that multiplicity in the name of abstract, universal concepts, such as 'Man'.

Science, in its idealized form, exerts the same sort of violence on technology as philosophy exerts on politics. Science tends to reduce the multiplicity of artifacts, of construction stories and contingencies to a *unicum*, to a unifying theory, to a minimal set of universal and eternal laws. Under the influence of philosophy, history has become historicism and has tried to cancel out all contingency and to annul the creative spontaneousness of politics and praxis, as much as science has neglected every particular event to consider important only what is universal, objective, and repeatable. But history is made of single, particular events: the thrust of historical dynamics stems from singularities. And these particular, singular events have to be narrated: to narrate means to recognize the contingency that holds sway over active life.

Historians rewrite the past again and again, because the past remains open and never elapses. The past continues to need justice and interpretation, history is not finished, is not transparent to itself. The objective sources of history, documents, are only one part of the basis on which historians work, thc othcr parts bcing narration and poctical invention.

Actually not only biographies and events have to be narrated, but also tales, novels and fictions stories, that are able to illuminate aspects of reality that would otherwise remain obscure since they cannot be read according to the criteria of traditional historiography. Novels and stories are sometimes able to comprehend the new, unprecedented events in a more subtle and complex way than the historical reconstructions. I am not saying that literature can or should replace historiography, but it can sometimes 'comprehend without justifying', as Hannah Arendt said [Guaraldo, 2003]. The true historical knowledge is based on a mixture of documents and rhetoric, and the two should go together. With respect to the comprehension of the unprecedented, the difference between documents and narrative vanishes. In fact, if reality is ununderstandable and unrepresentable, only fiction can represent it. Fiction offers itself as a means to understand politics through the imaginative liberty of literature and reverses the traditional supremacy of theory over practice, of mind over body.

This is precisely what technology is doing with respect to science. Technology is the ensemble of practical, or constructive, tales (machines, tools, systems, devices), whereas science is the ensemble of abstract laws. Technology, like narration, brings out the local, singular aspects as opposed to the global and universal ones considered by science. It brings out the bottom-up, step-by-step knowledge as opposed to the knowledge based on the abstract, Archimedean point of view (in this context *bricolage* is of paramount importance).

Technology exhibits the contingency of its constructions, hence sheds light on the provisional character of the principles and fundamentals of science. In the same way, the biological creatures produced by evolution reveal the provisional character of our evolutionary theories. Politics, as well as technology and narration, is the place of what is questionable, corrigible, modifiable. Politics, as well as technology and narration, opens up a region that eludes systematic order and abstraction, a region full of novelties, unexpectedness and creativity. Through narration, history as well as biological history, acquire more than one possible outcome, open up to context and eventfulness, intertwines with contingency and randomness. Thus humans regain a central position and the 'what is' of being is replaced by the 'who is' of acting [Guaraldo, 2003].

In the same way, the unifilar and progressive abstractness of science is replaced by the living history of technological artifacts, results and inventions. The products of technology are negotiable and perfectible, we never talk of the 'perfect, final machine', while the aspiration of science is always absolute, science always tends to the construction of a final theory that explains everything once and for all. To the cultural imperialism typical of western culture, science adds an epistemological imperialism that relies on Truth and Method to impose a unique model of thinking, crushing all other forms of thought and knowledge.

7 Conclusions

In the classical scientific vision, the universe is the stage of identical repetition, in which only a statistical fluctuation of tiny probability has started history. Life is the result of a sequence of repeated miracles eluding the fundamental determinism that would forbid all novelties and creations. Due to the necessity and inexorability of physical laws, the origin of life is a phenomenon about which science has nothing to say because of its uniqueness. In the new, emerging vision, however, the roots of life seem to plunge into an irreducible tangle of contingency and necessity. In this vision, singularities are the really interesting phenomena. In studying (or better in playing) this infinite play that is the universe, we should not consider only regularities and invariants, rather we should focus also on creative, exceptional phenomena, i. e. on single events. Thus, initial conditions, symmetry breakings, irrevocable choices, observers' viewpoints… should all be parts of the narration. History has burst into all sciences and all sciences have questioned the traditional image of history. Apparently, this indicates that also the traditional view based on the supremacy of being on becoming is a historical contingency: from Parmenides and Plato onwards, the concept of eternal and atemporal *forms* has concealed the *processes* that generate those forms. As a consequence, western philosophy has reduced itself to an ceaseless commentary to Plato. We can interpret all this by saying that a historical contingency (or metacontingency), i. e. Plato's existence, has brought about the development of another contingency, situated at a different level, i. e. the 'platonic' nature of western culture and science.

The universe of possibilities regenerates itself ceaselessly, in a discontinuous and unpredictable way, since the ensemble of possibilities is not given a priori once and for all. Modern science brings in itself the indelible mark of the age in which it was first born, and thus reveals its irremediable contingency and its inextricable entanglement with all other histories that the world tells about itself. And this narration is winding, intricate and arborescent, confirming that history does not arrange its events, thresholds and turns in a linear, continuous sequence allowing us to explain why and how happened what happened,

i. e. what, in this perspective, was inevitable. In fact, in the origin, development and interlacement of all histories there was no inevitable necessity. But these histories have happened, irreversibly, and it is in and by their interweaving that the present state of the universe, of life, of species, of civilisations has been created.

Bibliography

P. Bichsel, *Il lettore, il narrare*, Marcos y Marcos, Milano 1989.
O. Guaraldo, *Politica e racconto*, Meltemi, Roma 2003.
G. O. Longo, *Homo technologicus*, Meltemi, Roma, 2001.
G. O. Longo, *Il simbionte*, Meltemi, Roma, 2003.
G. O. Longo, 'Il velo oscuro: scienza e narrazione', in G. Lanzavecchia e M. Negrotti (eds.), *La scienza e la parola*, Scheiwiller, Milano 2004.

Giuseppe O. Longo, Dip. di Elettrotecnica Elettronica Informatica, Università di Trieste

Victor Margolin

The mechanization of exchange

Abstract:

Historically, interaction with others has been the principal way that humans have learned to participate in a shared social experience. At the core of this process is human exchange. The introduction of mechanized exchange is motivated by three factors: a) cost reduction; b) convenience; and c) efficiency. Mechanized exchange differs from human exchange in several distinct ways. It is less flexible, without affect or feeling. and does not allow for spontaneous engagement with another person. The rapid spread of mechanized exchange will surely encourage us to treat others more operationally. To the degree that we mechanize our relations with others, we will radically alter the socialization process and deprive ourselves of the humanity that has historically been our birthright.

Keywords:
human exchange, mechanized exchange, social transaction, intersubjectivity

1 Introduction

In the 1880s, Londoners witnessed the inauguration of the first vending machines. Initially, the machines dispensed postcards at a modest price but subsequently they offered a widened range of goods that included books, cigars, stamps, soft drinks, chewing gum, sanitary napkins and hundreds of other items. Although the vending machine may appear to have been a far less consequential invention than the telegraph or telephone, it introduced a powerful new concept to the realm of social life – the mechanized exchange. Now one could purchase goods from a machine without any human intervention.

For more than a century, inventors and designers have devised many other types of exchange that have substituted a machine or mechanized process for a human being. Of these, the first vending machine was the most primitive. Even though later machines could deliver coffee that was heated or soft drinks that were cooled, they still paled before the complexities of more advanced mechanized devices such as automated telephone switching systems that could substitute for a human being in the equivalent of a verbal exchange rather than a simple transfer of goods. The most ambitious proponents of mechanized exchange have been researchers in the field of artificial intelligence who believe that machines may one day take over much of the mental work that humans now do. They have devised expert systems to counsel human beings on medical ailments, financial investments, and military decisions. One of their crowning achievements is a program that enabled a computer to compete successfully against chess champion Gary Kasparov. The pervasiveness of mechanized exchange in contemporary societies raises questions about its effect on traditional social relations; notably its increasing role in enabling human beings to avoid contact with each other and its characterization of social exchange as a process dedicated to the accomplishment of specific goals.

Historically, interaction with others has been the principal way that humans have learned to participate in a shared social experience. Good citizenship is based on the capacity to work together with other people, whether in a neighborhood, an institution, a nation, or a global movement. The stages of human development, as they have been articulated by psychologists such as Erik Erikson are characterized by a move outward from the world of the self to that of others. At every developmental stage, the infant, the child, the young adult, and the grown person must learn to take his or her place in a family, a community, and ultimately in the wider world. Movement through these developmental stages includes learning to engage with others through participation in play groups, schools, work places, and social organizations.

At the core of this process is human exchange. Children develop a repertoire of social roles by relating to the people around them. They reinforce these roles as they grow up and continue to participate in situations where they learn new behavior through ongoing interactions with others. Alfred Schutz, whose project was the development of a phenomenological sociology, bases his theory of intersubjective social relations on a norm of mutual recognition. 'We are simply born into a world of others', he writes, 'and as long as we stick to the natural attitude we have no doubt that intelligent fellow-men do exist'[1]. For Schutz, humans are given to each other as counter-subjects who share 'a communicative common environment'[2]. They come to recognize the worth and value of their fellows through shared experience in which the exchange of speech and action is essential. A true social relationship, claims Schutz, requires a reciprocal awareness. I use Schutz as a reference because he emphasizes the role of intersubjectivity in the construction of social life. Other social theorists downplay this phenomenon but I side with Schutz in believing that the relations between people are as important as the cultivation of the individual self.

If intersubjective awareness is developed through participation in positive social exchange, then we need to consider the ways in which societies are structured in order to foster or inhibit such exchange. In the design of cities whose spaces derive from those of Greek and Roman urban forms, the best public spaces support the idea of common experience. The Greek agora, for example, was a space where people met to talk and socialize; likewise, commerce and sociality were conjoined in the market square of the medieval town[3].

Before the invention of the telephone, people could only interact with those in their immediate presence. The telephone made it possible to build communities and networks that extended across time and space. It enabled people to divide their consciousness between family, friends, and associates who were part of their physical environment and those who were elsewhere. Similarly with mechanized exchanges,

1 A. Schutz, 'Interactional Relationships', in A. Schutz, *On Phenomenology and Social Relations: Selected Writings*, edited and with an Introduction by H. R. Wagner, The University of Chicago Press, Chicago and London 1970, p. 163.

2 *Ibid.*, p. 165.

3 For a discussion of these spaces in the context of urban history, see L. Mumford, *The City in History: Its Origins, Its Transformations, and Its Prospects*, Harcourt, Brace & World, New York 1961.

the earliest ones required the physical presence of a vending machine while today such exchanges can occur at a distance by telephone or via e-mail or other electronic means.

2 Norms for social exchange

Given the rate at which mechanized exchanges are replacing human ones in every field from commerce and law to education and medicine, it is important to consider how we can assess the value of these exchanges so as to better understand their effect on the way human beings organize and conduct their social relations. To do so, we require a norm for meaningful social exchange against which we can evaluate the social consequences of increased mechanization. In this paper, I will simply speculate on what that effect might be without engaging in ethnographic research. However, mechanized exchange is qualitatively different from human exchange and I want to argue that the difference between them may result in significant transformations of human socialization.

First, I will return to the sociology of Alfred Schutz as a basis for articulating norms of social exchange. Schutz bases his model of exchange on the mutual recognition of human worth. He articulates this belief as follows:

> [t]he fact that the Thou who confronts me is a fellow man and not a shadow on a movie screen – in other words that he has duration and consciousness – is something I discover by explicating my own lived experiences of him[4].

Here Schutz makes the important point that one's own lived experience is the basis for understanding others. A life that is rich in human exchange should enhance one's understanding of others, while a corollary to this statement is that the more impoverished our experience of such exchange, the less likely we are to understand others and recognize that their needs are similar to ours.

4 A. Schutz, 'Interactional Relationships', *Op. Cit.*, p. 168.

Schutz raises another issue, that of co-presence. 'The face-to-face situation', he writes, 'presupposes then, an actual simultaneity with each other of two separate streams of consciousness'[5]. Given the widespread adoption of the cell-phone and the internet as intensive forms of human exchange, we have reason to question Schutz's insistence on the primacy of face-to-face communication and to recognize that exchange at a distance, whether by phone or electronic text transfer, has also become central to the experience of human social life for many people. However, the term 'absent presence' is now used to characterize the way that communication at a distance may compete with one's face-to-face relations and lead to rude or asocial behavior towards others in one's immediate physical environment.

If we identify mutual understanding and respect as ideal consequences of social relations, we can then consider the role of mechanized exchange in enhancing or detracting from them. The issue is complicated, however, by the fact that each person engages in many different exchanges each day, some mediated and some not. While face-to-face human exchange was once the dominant form of social relations, it is now joined by human exchange at a distance via the telephone or internet and mechanized exchange that enables people to obtain goods and services without the intervention of another human being. The question is not whether mediated communication is valuable or not in an absolute sense but rather how it can be balanced with face-to-face human exchange.

We must also recognize that not everyone recognizes the primacy or necessity of face-to-face communication and that some may even seek technological means to reduce it. A recent survey conducted by the National Restaurant Association in the United States found that seventy-five percent of consumers between the ages of 18 and 24 would prefer to place orders at a drive-in restaurant by using a touch screen rather than speaking to a human employee. Now restaurants are scrambling to install such screens, which have the added advantage of eliminating jobs and thereby reducing labor costs[6]. A discussion of this phenomenon must, however, be tempered by the fact that fast

5 *Ibid.*, p. 185.
6 'Fine, Have it Your Way', in *Newsweek*, April 19, 2004, p. 10.

food restaurants and big box stores such as Wal-Mart and K-Mart pay low wages to their sales people who may have little motivation for a meaningful social exchange with a customer.

3 A typology of mechanized exchange

Mechanized exchange can be divided into different categories based on three factors: a) whether the exchange occurs in physical space or at a distance, b) which medium is used, and c) whether goods or services are provided. Goods can be obtained either from a vending machine, a catalog, or a website. The vending machine is still a central device for dispensing goods through mechanical exchange in physical space. It substitutes for a human being as a purveyor of products, whether they be foods or drinks, train tickets, gasoline, cigarettes, or condoms. Other devices provide services in physical space such as access to or exit from a parking lot or a checkout facility at a supermarket.

Exchange at a distance can be through a physical object such as a mail-order catalog, or through an immaterial website. Instead of mailing an order to someone or talking to a sales person on the telephone, people buy goods on a website by following an electronic protocol that requires them to provide personal data in order to satisfy the conditions of the exchange.

When services rather than the sale of goods are offered at a distance through mechanized exchange, they usually consist of providing information or advice. Telephone companies, for example, employ voice recognition to answer requests for phone numbers. Likewise, airlines offer current information about flight arrivals and departures. More complex are the automated answering systems used by many companies that break down requests into separate categories and employ a combination of mechanized and human exchange to meet the user's needs.

4 Motivation for mechanized exchange

The introduction of mechanized exchange is motivated by three factors: a) cost reduction; b) convenience; and c) efficiency. These factors were central to the first wave of mechanization that was related to production in the 19th century. Inventors, especially in the United States, pioneered in creating machines that could substitute for human labor in the production process and this has led today to the automated factory[7]. The history of mechanized production has also been informed by the Scientific Management movement that was pioneered early in the 20th century by Frederick Winslow Taylor, who characterized workers as machines. Once the worker's physical movements were analyzed and broken down into separate operations, machines could then be designed to substitute for them[8]. From the employer's point of view, replacing a worker with a machine had many advantages. It guaranteed absolute obedience, it was cheaper, and it alleviated the requirement to consider the worker's needs and well being.

In the realm of production, manufacturers who adopted machines rarely had to face criticism from their customers. However, the automation of exchanges that affect customers directly is different. We therefore need to explore the reasons why a customer or user would prefer mechanized exchange or at least tolerate it. In most cases, customers have no choice. When a large company decides to install an automated telephone system, it simply does so and users must accept the consequences. There is no public debate or design criticism about such technological transitions. Companies consider the possibility of customer dissatisfaction and then weigh its cost against the savings they expect to achieve by providing a less expensive automated service.

Young people today do not even realize that live operators once answered telephones in offices and switched the caller directly to the

7 The politics of this process are discussed by D. Noble in his book *Forces of Production: A Social History of Industrial Automation*, Alfred A. Knopf, New York 1984.

8 F. W. Taylor, *The Principles of Scientific Management*, Norton, New York 1967, c. 1911.

person they wanted to speak with. If the caller was not sure whom to contact, the operator frequently knew. Sometimes one could strike up an informal conversation with an operator and experience the pleasure of a friendly or even flirtatious voice on the other end of the phone line. Characterizing the French love of flirtation, journalist Polly Platt writes that

> [f]lirting French means acknowledging the opposite sex in a nuanced, tender way, whether it is the grocer, the post office clerk or the concierge [...] French people marinate in a sort of subterranean erotic awareness all day. They do as much business as possible, whether in a bank or at a market, with the opposite sex[9].

Companies that do extensive business by telephone tend to establish a cost per transaction and then seek ways to maintain it. Whereas airline ticket agents once spent as much time as they needed with their customers in order to help them sort out the details of a complicated trip, today they are encouraged to enact scenarios with customers that drastically reduce the amount of time spent with them[10]. As an incentive to do away with ticket agents all together, the airlines offer lower fares on the internet to encourage people not to contact a live agent at all. They have also begun to add service charges to telephone transactions. What frequently happens is that companies initially wean customers away from live services by offering incentives or economics or convenience and then they discontinue the live services when they believe that a sufficient number of people will tolerate the loss. I see this process at work in my neighborhood supermarket where customers are encouraged to use automated checkout machines because it is quicker than waiting in line. However, to install the machines, the managers reduced the number of check out lines, thus insuring that customers would have to wait longer if they chose not to use the machine.

9 P. Platt, 'A Charm Course for France, from Paris...', *Chicago Tribune*, (Travel Section), April 18, 2004, p. 5.

10 B. Garson, *The Electronic Sweatshop: How Computers Are Transforming the Office of the Future into the Factory of the Past*, Simon and Schuster, New York 1988, pp. 40-70.

In fairness, one should note that automated systems are sometimes more efficient and desirable than other alternatives. In situations where there are high volumes of telephone calls, automated systems can put callers on hold in a line and insure that they will get through to someone eventually, even if they have to wait for an extended time. This is better than the days when one might have to try far too many times to reach a frequently called number because all the lines were tied up.

Similarly, on-line shopping can expose customers to a far greater choice of goods than shopping in a local store or even ordering from catalogs. It is now likely that a buyer can find almost any book he or she seeks, whether new or used, through one of the on-line websites that networks hundreds of small bookstores, which then have access to a worldwide market. On-line shopping can also be done around the clock so it becomes extremely convenient for many shoppers, who can buy a book or a new dining room table at three o'clock in the morning if they choose to do so. Thus efficiency, convenience, and expanded choice are all motives of the consumer to engage in one or another form of mechanized exchange.

5 Characteristics of mechanized exchange

Mechanized exchange differs from human exchange in several distinct ways. First, it is less flexible, Such exchange operate according to fixed protocols that are designed to accomplish particular ends. These protocols may offer minimal flexibility but ultimately they are governed by the predetermined purpose of the transaction. Efficient at best, they are offensive at worst, particularly when there is a glitch in the process and no avenue for seeking help is presented.

Second, mechanized exchange is without affect or feeling. Its purpose is simply to secure a result as efficiently as possible. Some companies such as airlines that use automated telephone systems conceal their protocols with human voices that assume colloquial styles and suggest a spontaneous exchange. However, the fact that these voices are simply recordings only reinforces the difference between the mechanical exchange and a real human conversation. The attempt to introduce human affect into the exchange is simply a deception intended to alleviate the feelings of isolation that can sometimes be a by-product of mechanized exchange.

Third, mechanical exchanges do not allow for spontaneous engagement with another person. When talking with another human being, whether in the process of purchasing an airplane ticket over the phone or buying meat at a neighborhood butcher shop, the possibility for a conversation that is unrelated to the business at hand always exists. Sometimes such conversations can be purely social or they may lead to an exchange of information related to another transaction. Either way, they offer an experience of human communication that helps to reinforce the feelings of mutuality that Alfred Schutz describes. Bereft of genuine affect and the possibility for human spontaneity, mechanical exchanges become purely operational. Their only purpose is to achieve a predetermined goal.

6 Conclusion

Having characterized mechanized exchange as rigid and goal-oriented, I will move to the question with which I began this paper: Is the increase of such exchanges in contemporary societies leading to a significant transformation of human relations? While ethnographic research would be required to answer this question in a persuasive way, I will nonetheless suggest that the characteristics of mechanized exchange may become more indicative of social relations in general as such exchange becomes more widespread. Today, the volume of mechanized exchange is growing and usurping many of the social interactions in which human beings once engaged with each other. Social mechanization will continue to expand as companies automate both sales and services to cut costs and reach larger markets. Once a given social level of tolerance for mechanized exchange exists, it becomes all the easier for it to proliferate. What is happening now is a mechanization of consumption that is equivalent to the 19th century mechanization of production Siegfried Giedion described in *Mechanization Takes Command*[11].

The consequence of this mechanization that I am most concerned about is the growth of a quality that I will call 'operationalism'. It is defined by the focus on a result rather than a process and may encourage someone to treat human exchanges like mechanical ones. The quest for efficiency in transactions is in part the result of busy schedules and a shortage of time to enjoy the spontaneous quality of a human exchange that is unrelated to achieving a particular goal. Mechanical exchange reinforces goal orientation because it offers nothing else. As a social consequence it can result in a lack of interest in other human beings with whom one has to interact socially including merchants, bus drivers, service workers, and even colleagues. Operationalism, which is a material quality, corresponds to the lowest level of human relations that Martin Buber described in his seminal work *I and Thou*. In that book he distinguished between recognizing the most spiritual qualities in another person, which identify them as Thou, and reducing others to the equivalent of material things, which Buber sees as treating them as an 'It'[12]. The rapid spread of mechanized exchange will surely encourage us to treat others more operationally. To the degree that we mechanize our relations with others, we will radically alter the socialization process and deprive ourselves of the humanity that has historically been our birthright.

Bibliography

B. Garson, *The Electronic Sweatshop: How Computers Are Transforming the Office of the Future into the Factory of the Past*, Simon and Schuster, New York 1988.

S. Giedion, *Mechanization Takes Command: A Contribution to Anonymous History*, Oxford University Press, New York 1948.

11 S. Giedion, *Mechanization Takes Command: A Contribution to Anonymous History*, Oxford University Press, New York 1948.

12 M. Buber, *I and Thou*, Scribner, New York 1958. For a relation of Buber's ideas to the practice of graphic design, see V. Margolin, 'Community and the Graphic Designer', in *ICOGRAPHIC*, 2/4, February 1984, pp. 2-3.

D. Noble, *Forces of Production: A Social History of Industrial Automation*, Alfred A. Knopf, New York 1984.

A. Schutz, *On Phenomenology and Social Relations: Selected Writings*, edited and with an Introduction by H. R. Wagner, The University of Chicago Press, Chicago and London 1970.

F. W. Taylor, *The Principles of Scientific Management*, Norton, New York 1967, c. 1911.

Victor Margolin, Department of Art History, University of Illinois at Chicago

Giuseppe Padovani

The user of the sites of the Italian *Comuni*

Abstract:

When the administration of a Comune *puts in the Web an hyper textual product, this produces a signification effect, since it offers both a representation of its own city management policy, and a representation of the addressee/user. The analysis of Italian* Comuni *sites singled out essentially two types of relationship between the administration and the citizen. One type of relationship privileges a host of services offered to the user/ citizen, while the other is centred upon the urban identity of the city and addresses to the citizen in general.*

Keywords:
public communication; the internet; semiotics

1 Introduction

In recent years, the diffusion of new media has enabled authorities and public institutions – particularly administrative and communal ones – increasingly to transfer to the web a wide variety of information and services in the form of portals and sites addressed to the public of users. This innovation has led the administrations of Italian *Comuni*[*] to use new forms of communication, that inevitably bring about changes in policies aimed at urban communities. In this perspective, joining the web does not represent a simple opportunity; rather, it constitutes *per se* a functional and symbolic enhancement of the product that is trusted to the web, and of urban government policies. It produces a

* In the present research we have used the home pages of Italian *Comuni* present in the period September-December 2000.

signification effect: it sets up a semiotic process. In addition, it modifies behaviour. For example, knowing that by connecting to the site of a *comune* one can find registrations of births, marriages and deaths through which documents can be printed out directly, or that the digital signature allows the transmission of legally valid documents, leads to a change in the behaviour of the user, and transforms the latter into a new type of user, active and not passive. At the same time, connection to a site brings about a modification of the behaviour of communal administrations who are able to put these services directly at the user's disposal, thus decentralizing them. By changing the behaviour of the users, the Internet in a sense constitutes an unlimited catchment area, as it were.

The connection through the mediation of the Internet leads to a new alliance between the *Comune* and the citizen: the medium enables an administration to enhance its policies by transmitting a representation of their action which would otherwise be impossible. Indeed, when the administration builds a hypertextual system in which they make themselves apparent, the web mediates between administration and citizen.

2 The site of a text

The arguments that we shall develop in the present paper move from an analysis of some aspects of the meaning-generating process in textual communication. This process does not consist in analyzing separately the communication poles (sender/receiver; addresser/addressee), or the message contents, independently of the agents involved. In fact, the process of generating the meaning of the text is given in the text, and lies in the relation between the poles. The text (we regard the portal and/or the site as a text – a communication artifact – 'is the place where the competences and practices of the addresser/addressee are realized and actualized' [Savarese, 1991, p. 188]. The text is a function of the communication for which it is produced. We can repeat with Umberto Eco that a text 'is a lazy machine that demands a fierce cooperative work' [Eco, 1979, p. 25]; 'a text is a product whose interpretative destiny must be part of one's own generation mechanism' [Eco, 1979, p. 54].

We think that a text – a portal or a site – actuates a 'strategy' by which addresser/ addressee cooperation is not limited to the varied and diversified modes of production and fruition – to what in a text becomes possible for the interpretation, but rather it involves a specific activity: that the text be used. The text demands to be used and the medium enables the text to be used. The text itself becomes a place for practical action, for use, *e.g.* a place where it facilitates the production of other texts. In other words, the relation between the sender and the receiver, that is incorporated in the text, when it addresses the addressee's competence and initiative, appears to be shifted towards the latter. In turn the addressee, according to his own interests, has the possibility of using the text. It is the medium, or the computer, that transforms the model of the reader – the interpreter – in the model of the user – the utilizer. The medium transforms the reader into a user and as a consequence the text/home site in a type of use. The medium makes available to the reader/user a wide range of information and services that he can use directly, according to his abilities and interests, without any mediation by offices or other persons specifically assigned to this task. That is, these new technologies enable the user to directly perform the operations whose execution would otherwise be assigned to specific persons – *e.g.* getting money from a cash dispenser, writing out certificates, etc.

Italian *Comuni* have understood the role of the Internet in public communication. As we said before, by constructing their own site on the Internet, the *Comuni* are able to reconcile a wide range of objectives: the public objective, aiming at making the *Comuni* known, to give useful and easily traceable information, to facilitate the objective of making themselves known and to publicise their own municipal, cultural, and historical values, and to construct identification symbols; to reconcile the objective of supplying direct and autonomously usable services of use to the citizen through the offer of many spaces or places of direct communication, through which the Commune is able to learn the citizens' opinions, and so on. However, when the *Comuni* design a home page that offers a host of on-line services, they give a representation of how they perceive and view themselves, and of how they intend to offer themselves to those who navigate, to the others in the site. In providing an image of themselves, the *Comuni* inevitably send an image of the other to whom they address.

More specifically: a medium enables the user to display a text. The virtual text exhibits a visible structure generated by an invisible one, that confers peculiar and distinctive characteristics to the first [Bettini, 1991; Colombo, 1993; Bettini-Colombo, 1993]. It is a cultural product. The *Comuni*, putting this cultural product on line, express themselves in a social communication. They address themselves to a potential user who we can define as a 'model user' on the analogy of the 'model reader'. Although the *Comuni* are aware of the gap between the user simulated in the text, and the real user entering the site, they try to capture the user. Similarly, the navigator, who visualizes the home page on the screen of his computer is able to form an image of the administration. These two aspects are interlinked. On the one hand, the identity of an artifact that can be qualified as a specific product – the *Comune* home page; on the other, the identity of the navigator that can be qualified as a user. Thus, a virtual text organizes one and the same textual structure:

a. The configurations organizing the representations of the sender. What institutional representations are generated by the site? Which are the modes that govern identity construction?
b. The configurations that organize the receiver's representations. What type of relations are possible with the user? What type of receiver is the site addressed to?
c. The configurations that organize and structure on line services. What subjects, services, etc. are treated in the site?
d. The configurations that organize the operation of the program and of the text.

3 Constructing a *Comune* site

In setting up a site, the discourse must meet two conditions. The first is a standard condition and concerns the more or less consolidated practice of programming – problems of accessibility, usability and functionality (*e.g.* in the computer the monitor always has significant limits and the author cannot know in advance what type of monitor will be used [Nielsen, 2000, 2002; Norman, 1990, 1995; Postai, 2003], The second is 'poietic' and privileges the graphic and aesthetic aspects of the communication artifact. These conditions are, on the one hand, opposed to one another: those governing the architecture of the site, standardized and normative in nature, are practical and technical; the second ones – the instances that refer to the form the text must have – are original and innovative. On the other hand, these conditions must cooperate in putting on line one and the same artifact. Their weight in the text can be different. However, both reveal the emergence of diversified signification traits. The site of a *Comune* must comply not only with usability needs, but must exhibit characteristics of recognizability and identity.

The monitor of a computer is a frame, given in advance, in which the programmer must in turn build a second frame. The space of the screen represents a sort of white sheet (that has to be paged up, or squared, if spaces are to be singled out, and where it will be possible to put 'objects' (a mix of objects, in our case). These objects are placed in the spaces and are highlighted by the paging up of the frame that separates at the same time one space from another. Then, division and increase of the spaces in the frame is carried out following the order of the objects. The order is an expression of the desired form, and of the meaning one intends to communicate. The order orients the spaces in which the screen is divided, and regulates the operations according to several axes – horizontal, vertical, of presence and depth: high/low, left/right, up/down, etc.

But the paging is not essential. The objects may be arranged at random by taking the monitor as a sort of table where the objects are placed either following one's inspiration or a scheme fixed in advance.

The paging up can be of many types:

Type 1.	New paging
Type 2.	Prearranged paging, e.g. newspaper headings in its many variant forms
Type 3.	No paging, the spaces being filled at random, e.g. notice boards.

In the first case the screen, like a sheet of paper, is squared up by means of perpendicular lines that mark the spaces where the objects are to be placed. In the second case, the monitor is seen as a newspaper heading. A rectangle is marked at the top, and the space below is divided into vertical columns.

The paging marks the boundaries of the overall space and at the same time creates the spaces allocated to the objects. It forms a global composition, or scheme. It designs the general arrangement of the spaces. The design does not have a direct relation with the objects. It is an organized unit consisting of lines, points, colours and geometrical forms, and at the same time it attracts attention as an overall representation that belongs to the order of visibility and reading. More importantly, the graphic choice of the paging is not indifferent with respect to the type of message intended for the user since it sets up the contents and decides their orientation. From this point of view, the home page acts as a means of identification: the paging, the architecture that controls space organization – character choice, colour, images, placing of the links – contribute to the production of one representation, the 'visual identity' of the site.

Essentially, the home page is a 'graphic representation' or, to quote G. Anceschi's definition, a 'communication artifact'. It is a mix, rather than a hypertext, of schemes, frames, drawings, photographs, writings, at times moving images, accompanied by music and sounds. But it is a very peculiar type of mix, since it contains instructions that make it possible to work directly in and with it.

I want to emphasize the words 'artifact' and 'communication'. The first is reminiscent of the artificial/artificiality concept, meaning artifice: something made by man with skill and ingenuity [Padovani, 2000]. The second refers to the concept of communication. In this case, however, the word takes on a new meaning, since – as mentioned above – interactive operativeness is a constituent part of the same process, since it is embodied in it.

Both the computer and the artificial product enable the receiver of the communication to perform certain actions that make him an *active subject*, not simply a passive one. This aspect of the artifact reasserts the pragmatic dimension of communication. In fact, as recalls Lotman

[Lotman, 1993] – every communication act is at once a social, cultural and historical interaction and partakes of the depth of social memory.

The first page of a site undoubtedly falls within the frame of visual communication and can be analyzed in this perspective. But it is a very particular product – a cognitive artifact – since it offers itself not only to the sight, or reading, but also to the touch, or manipulation. Because the message/text does not perform exclusively the function of referring to, the addressee, in addition to receiving the message/text, acts in and with it.

For example a single word often serves as a whole sentence. It has a performative value. It is an invitation, an order, a promise: one can point at and click on it with the mouse. By opening a new page it opens up a new scene. It makes visible what is distant and absent. The word goes along with an arrow, a button, a sign that makes it differ from an ordinary word, transforming it in a sort of pictogram or ideogram [Cardona, 1981]. The word, or the button to be clicked on, has an aura of magic. In addition, these 'words' have a strategic position in the economy of the graphic representation. Words, images and drawings are staged, put in a painting, and meant not only as an image representing something to someone, but also as signs that will make someone do something. Their function is that of indices: they make relative the situation/context of the enunciation, making it interactive to the addressee.

What must a site of a *Comune* represent? A *Comune* site cannot but 'communicate' a 'message' that proceeds along a continuous line at whose sides we can find either the *Comune* as a municipal/administrative/political centre or the *Comune* as an urban/economic/cultural centre. The two poles of the line enhance the two dimensions of the city. One site privileges one or the other of the poles without giving up different degrees of mixture.

In any case the designer of the home page must devise and work out an image of the city, a recognizable symbolic representation in front of an everchanging reality (information, online services, etc.). The image must be anchored to a project, to a representation, at the same time comprehensive and synthetic (after all the home page is a form of anchorage, a point of arrival, departure and homing. It is the home, the den or the home-base of baseball).

The staging of this image can be conceived like the preparation of newspaper heading. In the case of a newspaper the heading presents a permanent graphic design, while the news, comments, etc. change from day to day. Further, a newspaper follows an editorial policy, and has an ideological orientation that directs the choice of the news and how it is reported. Similarly, a portal/site selects, isolates and separates from the universe of phenomena and events concerning the life of the city, certain events, news, reportages, and connections. These take into account the editorial policy and of the intended image of the city.

In fact, the point of view of the graphic designer 'implies a context of a host of information for elaboration, but above all it foresees a *spectator* to be reached with a host of expedients and devices of exposition' [Anceschi, 1992, p. 56]. Thus, the spectator, while selecting the news and elaborating the image, prepares a particular representation of the possible spectator/user. What is to be chosen (and consequently what is to be omitted) inevitably leads to the transmission of a privileged image of the users.

Since the representation the site proposes is the result of a selection/reduction process, it is a schematization. This process has an unavoidable dimension, and cannot be avoided Therefore it is a constituent part of the representation itself as well as of the signification process generated by the communication artifact. In this sense, the representation is an exhibiting (*Darstellung*). As Anceschi affirms, planning is more similar to the gesture: 'In any case, it involves an act, a gesture, that transforms an object into a bulletin'. [Anceschi, 1992, p. 25]. This decision, this gesture of exhibiting serves as an appeal, and has a *basic function* – the *anaphoric function* as Kristeva says, or more simply a *phatic function* as Jackobson claims. Whatever the medium, the gesture of pointing at – the indicating function – lies in front and behind the voice and the writing. However, any medium will inevitably take up again the voice and the writing, and aims at calling, at attracting the attention, at establishing contact. In our perspective, the home page has this specific function. This is the main reason that guides the analysis of the home pages of Italian *Comuni*.

The home page 'stimulates' the citizen/user; it appeals to his interests, to his cognitive dimension, but to his sensibility, culture and curiosity as well, to his aesthetic sense. The home page representation enhances perception and sensibility; through a mix of images, drawings and words, it acts as a stimulus. It can address a user for whom usefulness does not exclude other modes of fruition.

To sum up, the analysis of the home pages of Italian *Comuni* has revealed the existence of a conceptual axis on the sides of which we can picture two extremes. On one side we observe such values as convenience, practical aspects and usability, for the user. The user is mostly intended as a fellow citizen who must at once be able to find everything that an administration has to offer. The majority of the home pages prefer an architecture like the one adopted in newspaper headings (Fig. 1. *Comune* of Cuneo). On the other side we find the symbols of urban identity, and the emphasis is on such aspects as historical memory, monuments, cultural and social traditions. For the most part, these sites exhibit an original architecture, not classifiable under the newspaper heading type (Fig. 2. *Comune* of Sondrio).

Acknowledgment

The author gratefully acknowledges the help of Doctor Romano Zanni in translating from Italian to English. All errors, inaccuracies or misunderstandings remain, however, the responsibility of the author.

Bibliography

G. Anceschi, *L'oggetto della raffigurazione*, Etas Libri, Milano 1992.

G. Bettetini, *La simulazione visiva*, Bompiani, Milano 1991.

G. Bettetini and F. Colombo, *Le nuove tecnologie della comunicazione*, Bompiani, Milano 1993.

G. R. Cardona, *Antropologia della scrittura*, Loescher, Torino 1981.

U. Eco, *Lector in fabula*, Bompiani, Milano 1979.

J. Kristeva, *Séméiotiké. Recherches pour une sémanalyse*, Seuil, Paris 1969.

J. M. Lotman, *La cultura e l'esplosione*, It. transl., Feltrinelli, Milano 1993.

J. Nielsen, *Designing Web Usability*, New Riders, Indianapolis 2000, It. transl., *Web Usability*, Apogeo, Milano 2000.

J. Nielsen, *Homepage Usability*, New Riders, Indianapolis, 2002, It. transl., *Homepage Usability*, Apogeo, Milano 2002.

D. Norman, *The Psychology of Everyday Things*, Basic Books, New York 1988, It. transl., *La caffettiera del masochista. Psicopatologia degli oggetti quotidiani*, Giunti, Firenze 1990.

D. Norman, *Turns Signals are the Facial Expressions of Automobiles*, W. W. Norton & Company, New York-London 1992, It. transl., *Lo sguardo delle macchine. Per una tecnologia dal volto umano*, Giunti, Firenze 1995.

G. Padovani, 'The Artificial Between Culture and Nature', in *Al&Society*, nn. 3-4, Springer-Verlag, London 2000, pp. 300-313.

S. Postai, *Siti che funzionano*, Hops Libri, Milano 2003.

RUR, *Le città digitali in Italia. Rapporto 1998*, Franco Angeli, Milano 1999.

RUR, *7° Rapporto. Le città digitali in Italia*, 2003, <http://www.rur.it>.

R. Savarese, 'Grafica quotidiana', in M. A. Bonfantini and A. Martone, (a cura di), *Specchi del senso: le semiotiche speciali*, Esi, Napoli 1991.

Giuseppe Padovani, Dipartimento di studi politici e sociali, Università di Parma

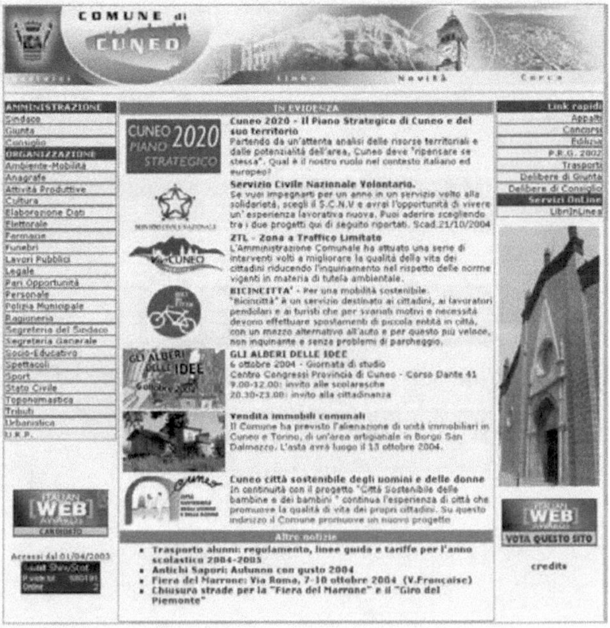

Fig. 1. *Comune* of Cuneo web site

Fig. 2. *Comune* of Sondrio web site

Antonino Porrello

Research of leading role of social sciences in the information and communication technologies planning

Abstract:

This work relates some theoretical and methodological considerations about approaches at the planning and technological innovation sharing, that in last years, at the international level, have checked more attention and agreement. Such considerations are the outcome of the research activity at the beginning of the nineties, and then they have found incentive and feeding in a series of recent discussions in some cycles of seminars organized by ASSOLab (at the Planning Department, IUAV University of Venice) with the participation of engineers, planners, computer scientists, sociologists and ethnologists.

The main theme faced was the analysis of implications and benefits resulting from a new technologies research that considers those as artefacts, and so that is able to catch both the planning function behind their implementation and the importance that this planning has to promote all innovation and social-organizational development activities.

Nowadays public opinions recognized without doubt that technologies are indivisible from their practical use and that there is a close connection between working context and activities organization: technologies and objects haven't a precise role, but their use is always contextual, depending on importance assigned by actors at the singular situation. Then, everytime someone uses a technological artefact leave a mark for the future action: so, during the time, technological objects are considered from actors as more or less considerable or contextual, negotiable or resistant.

Studies by ASSOLab in '90 years are deepening of research approach that conceives in the best way technologies as activities, showing how in contemporary working context human and technological resources are not in opposition, but they originate a representation ecology *that we can define also as* practices ecology. *Beings, symbols and machines, come together to structure and renew the comprehension of a social and organizing situation, while the good analysis and comprehension of dynamism of those situations can support and extend the work practices natural evolution, in right innovation processes.*

In the first paragraph we explain theoretical origins and some applicative developments of Participatory Planning approach, in the technological and organizing innovation context, because of the relevance of those studies in the last years, to support and manage organization changes. Then, we report a deepening about some qualitative methodologies, inspired by ethnography, that, in coherence with participative planning principles, permit to make original studies in organizing contexts, giving the right importance to situation complexity and to analyzed working processes.

In the paragraph number 4, we describe which theoretical-methodological considerations can be deduced from the type of research we have presented, to support the articulated intervention of user-balanced planning, that ASSOLab has developed in those last years to sustain innovation processes in productive ambient and in particular in the shoes manufacturing sector.

Finally, there are some reflection entries and proposals about future developments of this theme, that can be interesting for research and didactic activities actually treated in ASSOLab and in Planning Department (IUAV University of Venice).

Keywords:
accountability, design by doing, knowledge exchange, new technologies, participatory design, planning for changes practice, planning for during use planning, practices ecology, technological and organization innovation, users participation, users and planners

1 Participatory Design role in the technological and organizing innovation: theoretical origin and applicative developments

In the last ten years, the Participatory Design approach have represented the most diffused and innovative methodology at international level, to plan new information technologies, in particular the new informative and communication systems at distance. The peculiarity of this approach is the introduction of a new principle: persons doomed to use a determined technological artefact must play a crucial role also in the planning of this same system [Elm and Kyng, 1991; Ehn and Sjoergen, 1991; Ehn 1989; Shuler and Namioka, 1993; Kensing and Blomberg, 1998; Jacucci, 2001].

The Participatory Design (DP) represents a radical perspective change respect previous planning methodologies: we can remind, for example, the Business Process Re-engineering, that preferred business and organizing restructuring interventions, in which few managers, in exclusive, had the responsibility to direct the change process, to give more efficiency to an organization.

The principal objective of DP in every technological innovation process is to try to give workers the best instruments to carry out their task. So, informative systems are not considered isolated as regards the working context in which they are included, and they're not ended products but growing up processes: in this way, we can derive the concept of *in use design* or *in design in doing* [Schuler and Namioka, 1993; Henderson and King, 1991].

Theoretical origins of DP date back to Scandinavian tradition of studies and projects, begun in the seventies, to make more democratic the working context through the direct involvement of users in the technological development of organizations [Ehn and Kyng, 1987]. In the eighties, this methodology was applied to symbolic projects, as 'Utopia', in which for the first time the trade unions and management involvement in the develop of business informative network was experimented [Ehn, 1989]. In following ten years, the Scandinavian tradition interlaced American emergent experiences in planning cooperation of informative systems [Greebaum and Kyng, 1991] slowly losing its 'democratic' connotation and orientation, more developing pragmatic and applicative aspects in 'design in site' or in the 'contextual design'.

Two important aspects characterize the DP principles implementation: *Mutual Learning* and *design by doing*. The first is a mutual learning form between users and planners, through information tradeoffs as regards both the working practices and the technological possibilities available, in a context of activities end experiences carried out at the same time. This reflects the endeavour to establish a dialectical relation between languages and interpretative visions of users and planners, between need of social organization and requirements of technological systems to develop, to promote an useful knowledge exchange between involved actors. So, from an epistemological point of view, the DP becomes a chance to gradually and diffusely build the necessary knowledge to develop and use a new organizing artefact [Kensing and Blomberg, 1998].

Obviously, there is the possibility of misunderstanding and incomprehensions between participants of this process: this step must be faced, managed and mediate trying to establish an effective and open communicative context.

Instead, the second important aspect of DP, the *design by doing*, refers to all interactive processes of experimentation, modelling and evaluation, supporting project and learning activity. We can remind the most diffuse, *mock-ups* and prototypes iterative realization. *Mock-ups* are artefact realized to represent one or more system functionalities to develop, before its effective implementation or prototype setting-up. Materials that compose them can be disparate, sometimes are very simples and economics as paper, because of their utility consists to supply an idea of functionality type or an interface we want to realize, to allow a dialogue and a comparison which lead more quickly to following implementation phases of prototypes or of the final product.

The situated nature of each DP intervention has rendered this methodology widely recognized as the most adapted and effective to face challenges and complexities derived by new technologies adoption and development in today's productive context. In countries that first have used this approach in the innovation technology, planners and users of new technological artefact distinction aims to disappear.

2 Qualitative methodologies

We believe useful, in the following section, to examine closely the analysis of some qualitative methodologies that, starting from DP assumption, can permit to the researcher to orient himself better respect to observation-intervention modalities to adopt in organizing context, as also in more opportune instruments choosing for the management of this type of research.

2.1 Users Participation

The Participatory Design (DP), that was born in Scandinavian lands in the 70s-80s, have been revisited in those last years by a sociological point of view. In this way, the principal aim is to obtain, in the technology planning, a real users participation, to really resolve meeting problems in the use of the same technology. Actually, the participation problem is frequently studied, trying to give valid answers. The PD is, essentially, a learning process that requires long times. It's not only a requisites raising that a certain technology must have. Besides, those requisites change in time in relation to the technology use by users. Planners could consider those use changes. The success of a technology depends on its acquisition by users. If technology has been used in a different way respect the planner prevision, *Accountability*[1] arises. If there is the intention to create a technology going to users participation, it's necessary to take note of the accountability and to foresee, until starting project, the possibility that user changes the system configuration during its use, depending on his contingent necessities. The PD is particularly useful in informative structures planning for hospitals, law courts, etc., and so all companies that use technology to realize office practices.

1 The accountability shows a trend to change during use. Ciborra notes: 'By drifting I mean a slight or significant shift of the role and function in concrete situations of usage, that the technology is called to play, compared to the planned, pre-definite and assigned objectives and requirements, irrespective of who plans or defines them, users, sponsors, specialists, vendors or consultants'. 'Drifting can be looked at as the outcome of two intertwined processes. One is given by the openness of the technology, its plasticity in response to re-inventions carried out by users and specialists, who gradually learn to discover and exploit features and potentialities... On the other hand, there is the sheer unfolding of the actors' "being-in-the-workflow" and the continuous stream of interventions, *bricolages* and improvisations that "colour" the entire system life-cycle' [Ciborra, 1996].

2.2 Planning practices

In the sector of informative systems development, it goes forward a wide movement of new project practice request, a *planning for changes* practice and a *planning for during use planning*.

The Technology, that it's not technique, is the study and the rationalization, through science, of some different techniques. Through technology and science it becomes a production factor. Already in *Encyclopédie* (1751-1772) by Diderot and d'Alembert, and also in Carl Marx, the currently use of technology concept was referred to science application to productive processes.

In the Postindustrial society, a considerable and growing part of technology does not consist only in engines or other material machineries, but also in planning, programs, operative practices, organization models, decisional procedures, formulas for complex problems solving, so that technology concept necessarily hubs both technical praxis and its material and immaterial supports and vehicles.

It must be observed that there nearly isn't phenomenon of associated life that its origin or configuration was not attributed to technology. It's declared, in fact, that the modern technology has social consequences (it influences):

1. disguised as medicine technology on sanitary organization, on medical profession and on knowledge professionists and technicists, on patient condition;
2. disguised as productive processes technology, on work and company general organization, on professional structure and occupation;
3. disguised as informative technology, on work organization, on companies general structure, on companies management, on knowledge managers, professionists and technicists role;
4. disguised as decisional processes technology, strictly connected to informative technology, on giantism and direction of national and multinational companies [Gallino, 1978].

The contemporary technology development emphasizes social consequencies induced by technological change from technology user point of view, to *learning by using* to changes brought to an object not during the production, but in the context of its use by final users [Rosemberg, 1991]. This approach influences the professional structure of professionists and knowledge technics.

The technology changes with the use: users know very well their strategies, and also evolutions of those in time, so systems must evolve to remain effective. Although *tailoring* practice supplies an important contribute to a systems planning oriented to employment, the users planning during the use (PUU) appears as the ultimate purpose of actual movement.

This state of things recalls a similar phenomenon happened in architectural context during the last century: the research of a planning responsibility distribution by the 'open architecture' movement involved users in single building planning [Habraken, 2000]. Both in that case and in this one, the users capacity to during the use plan is obtained recording in artefacts and in organizations a strong change element: from the very beginning, the artefact planning must foresee the possibility of following modification of the project through the possibility of further planning by user during the use.

2.3 Open Participatory Design

The Participatory Design (PD) method, ignoring the Accountability and considering only technical aspects and not sociological ones of informative system, is not the best. A refounding proposal (Jacucci, 2001) of informative system planning could consist in supporting the abstraction (in which is founded the informative science, that abundantly resorts to symbolic language) to accountability. It's a sociological concept question, that explain why a technology is not used in all its potentialities. This happens, above all, for two reasons: because not all object functions are well-known, and also because it's not clear how those can be used. Accountability is the technology capacity to 'relate' itself, to explain its functions. It's a characteristic of whole made by planners and users group. Every user and planner give a different meaning to a technological object: it's a question to make a synthesis of those different meanings. Everyday a new technology is created, but just this innovation pressing induces planners to neglect Accountability, that instead it always was considered in the past, when technological evolution was slower and planning demanded longer times.

The change as fundamental element. In fact, users can rise to different level of comprehension and familiarity with informative artefacts. Therefore, technology accountability is the outcome of a learning process by organization, in which pre-existing work practices are supported by innovative instruments and new work practices are permitted by new instruments. Moreover, this is not either the only or more important change cause. On the one hand, organizations change in time because of internal context alterations, or as outcome of external pressures. On the other, information technology seems to change in a rising velocity on one's own. In conclusion, used in organizations informative systems must evolve to remain effective.

2.4 Planning the technology in a point of view of change and not of stability one

Technology change is imposed by accountability. Through an apposite interface, user can re-shape his informative system, modifying some precise internal aspects (algorithm) and leaving unvaried the external aspect, and so the relation with context in which system is inserted.

The technological object preserves the same external aspect, the same relation with other instruments connected to it, but inside changes depending how users employ it. Technology presents a hidden complexity and a surface one. Only the last is revealed to user. Through the technology predisposition to change, dialogue improves between users and planners, which normally use a very different language. The *Use Case* is an informative concept that permits the communication between users and planners. The use case of technology considers the context in which a technological object is used. Generally, planner describes only generic functions of a technology, abstracting it from specific contexts. Then, is the user task employing those functions in his particular context of use. At this level can manifest incomprehensions between user and planner, because the first can perceive to need definite applications that his informative system has not. Through the requisite picking that a system could have, the planner defines specific characters, which are considered in his project. Then, he submits to user a prototype, that anticipates the real final object, and that permits to user to understand what he needs to carry out his work. Before user finds himself meeting the object, or a prototype, he doesn't know exactly to want what then he'll asks.

3 User dedicated design approach in ASSOLab researches

In modern bearing systems is present, on the one hand a more emphasized use of technologies, on the other the science endeavours in order that technological innovations are more adequate to satisfy productive world requests.

Considering the shoes manufacturing sector, that in those last years is studied by ASSOLab, we can note that the introduction of technology in this productive system has generated significant innovations, contrarily to common opinion that hand-made plus-value is made above all by the presence of a strongly manual handicraft. Among those innovations we remember, in particular, carrying out software experimentations [Bucci, 1996] that permit to start the simulation process of the product.

The simulation, that is in experimentation in this period, in the shoe production context is of three-dimensional type, permitting the interactive intervention of the operator in real time. This technology innovation cleanly separates the planning part from the productive one, starting a new phase, in which a Participatory Design approach can easily became a part.

So, the *participating user* at this phase is a new professional figure which is inserted in the production context, with the exact intent to locate and formalize matters originated by the path from planning to final product realization. Therefore, the aim of ASSOLab researches is to obtain and underline, through a methodological design process adjusted on the user, which are the elaborated and used technologies in the shoe production context, and also the role of those technologies users.

In this connection are significant indications given by Bijker [1998]; in his opinion, the leading research must incline to:

- show the technological object flexibility and as the same artefact can be built in different ways and forms, because doesn't exist a unique optimal solution;
- analyze used mechanisms to close this interpretative flexibility, and through them object assumes a steady form;
- connect those closing mechanisms to widest social context.

Therefore, ASSOLab researches find to focalize functionality and type of participatory user intervention, considering how those aspects must answer expectations of change demanded in internal productive organization, still less to be involved as active element in the internal businness organization.

Participatory user, as well as must be considered as a consultant, also must provide the validity to the practical knowledge and must participate to a specific situation. Collaboration between the two parts (participatory user and the technician of production) takes the form of mutual learning among experts coming from different sectors, as previously underlined.

In the sporting shoe case, during the realization prototype phase, participatory user can be represented by athletes, which are involved (because of their ability and knowledge) in the quality examination of final product. Concerning this, we underline how is requested to them not only an opinion about the sporting shoe functionality, but also to notice problems to subject to technicians, which then are called to find a solution. Both of them concur to middle check of objectives and elaborate together the final product, that will become a consumption object by wider consumers groups.

Nevertheless, we retain that not always elaborated outcomes in planning and revision phase satisfy different uses asked by users/consumers. If from one hand we consider that not all artefacts has obtained the forecast success (often owing to inexact programming or to unclear objectives), on the other most of successes are due to a multilinear process and to a information exchange, in which participate not only planners and shoes producers, but also social groups of users as sportsmen, athletes and sport professionists.

Methodological research picture proposed by ASSOLab, tuning with research intervention principles, will try to favour an active co-operation between researchers and studied organization members, rendering the research a mutual learning process between the two parts. Infact, while researcher, in his mediator role, learns elements related to local and specific work practice, the local actor learns the practical meaning and the form of current innovation.

Research intervention trend will be spiral, including phases of:

1. starting negotiation of resources, roles and objectives;
2. work phase, as analysis and development;
3. valuation and new negotiation.

Besides, the intervention will be characterized by:

- focus on effective problems or change waiting expressed in the organization;
- involvement and active participation of all organization or a part of it;
- orientation to changes production and to 'knowledge in action' production on the sector;
- cyclicity of negotiation, analysis, implementation and valuation phases;
- interest in supply validity to local practical knowledge, connecting to the specific situation;
- vision of the researcher as consultant, and of the action as previous research;
- occasion of mutual learning among different sectors experts.

Principally, user will be called to participate because he's the only expert, even if tacitly, of his working practices. Then, it's suggested that he will better use an instrument that previously he has contributed to plan, converting the design aimed at user in a guarantee as regard the use of new technological artefacts.

However, the research intervention of ASSOLab will start from awareness that to create a real participation needs to define with care *who is* the user of the system to carry out, both technical (user) and political (manager) point of view, trying to negotiate spaces and resources to make easier a certain level of participation to decisions and planning, by both figures. So, it seems to be that, far from being a development delegacy to final user, this participatory design approach will ask a great obligation by ASSOLab researchers, to adequately interface organization actors to development practices.

From an organizing point of view will be defined working groups to carry out common activities, workshops to involve other organization members, executive councils to involve managerial elements. Groups and work plan generation will can be constantly the negotiation object, as also allocation of other necessaries resources to good research execution.

As regard workshops, which will be the principal instrument of participate intervention inside organization, will take care to negotiate with management the local actors participation at meetings, in which will come out the sense, contents, innovation evaluation in its phases and will be negotiated spaces and responsibilities of actions to achieve.

Participant's observations, interviews and other techniques of ethnographical inspiration will be used to generate learning and knowledge supporting the intervention, to render it more effective and shared.

Bibliography

Z. Barman, *La solitudine del cittadino globale*, Feltrinelli, Milano 2000.

W. E. Bijker, *Of Bicycles, Bakelites, and Bulbs*, The MIT Press, Cambridge (MA) 1995, It. Transl., Mc Graw-Hill Libri Italia, Milano, 1998.

K. Billeskow Boving and L. Hoffmann Petersen, 'Design for Dummies – Understanding Design Work in Virtual Workspaces', in T. Binder, J. Gregory and I. Wagner (eds.), *PDC 02 Proceedings of the Partecipatory Design Conference*, Computer Professionals for Social Responsibility, Malmo 2002.

J. Blomberg and F. Kensing, 'Preface to the Special Issue Partecipatory Design', in *Computer Supported Cooperative Work*, 7, 1998, pp. 163-165.

M. Bucci, *Scienza e società*, Il Mulino, Bologna 1996.

M. Castells, *The end of millennium*, Blackwell, Oxford 1998.

M. Castells, *La nascita della società in rete*, Università Bocconi, Milano 2002.

M. Castells, *Il potere dell'identità*, Egea, Milano 2003.

C. Ciborra, 'What does groupware mean?', in C. U. Ciborra (ed.), *Groupware and teamwork*, John Wiley, New York 1996.

C. Ciborra and G. F. Lanzara, *Labirinti dell'innovazione. Tecnologia, organizzazione, apprendimento*, RCS, Milano 2000.

P. Ehn, *Work-Oriented Design of Computer Artifacts*, Lawrence Erlbaum Associates, Hillsdale (NJ) 1989.

P. Ehn and M. Kyng, 'The Collective Resource Approach To system Design', in G. Bjerknes, P. Ehn and M. Kyng (eds.), *Computer and Democracy – A Scandinavian Challenge*, Avebury, Aldershot 1987, pp. 17-58.

P. Ehn and M. Kyng, 'Computer: Mocking-it-up or Hands-on the Future', in J. Greenbaum and M. Kyng (eds.), *Design at work: Cooperative design of computer systems*, Lawrence Erlbaum Associates, Hillsdale (NJ) 1991.

P. Ehn and D. Sjörgen, 'From System Description to Scripts for Action', in J. Greenbaum and M. Kyng (eds.), *Design at work: cooperative design of computer systems*, Lawrence Erlbaum Associates, Hillsdale (NJ) 1991.

L. Gallino, *Dizionario di sociologia*, UTET, Torino 1978, pp. 722-726.

G. Gobbo, *Descrivere il mondo*, Carocci, Roma 2002.

J. Greenbaum and M. Kyng (eds.), *Design at work: Cooperative Design of Computer Systems*, Lawrence Erlbaum Associates, Hillsdale (NJ) 1991.

N. J. Habraken, *De drags en de mensen*, Scheltema & Holkema N. V., Amsterdam 1961, It transl., *Strutture per una residenza alternativa*, Il Saggiatore, Milano 1974.

N. J. Habraken, *The Structure of the Ordinary*, The MIT Press, Cambridge (MA) 2000.

A. Henderson and M. Kyng, 'There's No Place Like Home: Continuing Design in Use', in J. Greenbaum, M. Kyng (eds.), *Design at work: Cooperative Design of Computer Systems*, Lawrence Erlbaum Associates, Hillsdale (NJ) 1991.

G. Jacucci and D. Calzà, *Uso di Cases nel progettare sistemi prevedendo la progettazione da parte degli Utenti durante l'Uso: un caso relativo a sistemi di supporto alle decisioni*, Seminario ASSOLab, Venezia 2001.

F. Kensing and J. Blomberg, 'Participatory Design: Issues and Concerns', in *Computer Supported Cooperative Work*, 7, Kluwer Academic Publishers, 1998, pp. 167-185.

M. Kyng and L. Mathiassen, *Computers and Design in Context*, The MIT Press Cambridge (MA) 1997.

S. Leith Star, 'Infrastructure and Ethnographyc Practice: Working on the Fringes', in *Scandinavian Journal Information System*, vol. 14, n. 2, 2002, Special Issue: 'Ethnography'.

S. Leith Star, 'The Ethnography of Infrastructure', in *American Behavioral Scientist*, vol. 43, n. 3, November/December 1999, pp. 377-391.

A. Porrello (eds.), *Modelli dell'utente nelle società avanzate*, Ed. IUAV, Venezia 2003.

A. Porrello and D. A. Collovini, *L'approccio partecipativo nella scienza e nella tecnologia dei sistemi simulativi*, Ed. IUAV, Venezia 2003.

N. Rosemberg, *Inside the Black Box, Technology and Economics*, Cambridge University Press, Cambridge 1982, It. transl., *Dietro la scatola nera*, Il Mulino, Bologna 1991.

D. Schuler and A. Namioka (eds.), *Partecipatory Design: Principles and Practices*, Lawrence Erlbaum Associates, Hillsdale (NJ) 1993.
A. Touraine, *Critique de la modernité*, Fayard, Paris 1992, It. transl., *Critica della Modernità*, Il Saggiatore, Milano 1993.
A. Touraine, *Libertà, eguaglianza, diversità*, Il Saggiatore, Milano 1999.
A. Touraine, *Comment sortir du libéralisme*, Fayard, Paris 1999, It. transl., *Come liberarsi del liberalismo*, Il Saggiatore, Milano 2000.

Antonino Porrello, IUAV, Venice Institute of Architecture, University of Venice

Barbara Rondelli

Learning communities in the demand for postgraduate education

Abstract:

Net learning favours acquisition of the competency required by organizational structures based on knowledge management, *and in societies that desire to widen the opportunities of* lifelong learning *for all. Net learning is a new way of producing knowledge. It requires the use of technology, of different methods of planning and of providing the educational process, different roles for teachers and students. Active participation by teachers and students is aided by creation of 'learning communities' that promote continuous interaction among users who become transformed from passive into active participants, in full awareness and responsibility for their ways of being and behaving. The user is here seen as an active person who interacts and shares expertise, knowledge, solutions and ideas. The user is not seen solely as a utilizer / beneficiary, but is also a contributor.*

Keywords:
learning, technology, cooperation/collaboration, competency and knowledge, innovation, interaction, support, user/active member

1. Introduction

Net learning facilitates acquisition of the skills required by organizational structures based on knowledge management and by those societies that aim at extending opportunities of lifelong learning to all. Net learning is a new way of creating knowledge. It involves the use of technology, a variety of systems for planning and carrying through the educational process as well as changes in the roles of teachers and students. The active participation of teachers and students is helped by establishment of what are called 'learning communities' that favour continuous interaction among users of educational services who are no longer passive but acquire awareness and responsibility for their modes of living and behaving. The various users of the educational process (*stakeholders*) take an active part in carrying the educational process forward, interacting and sharing expertise, knowledge, solutions, ideas, thus also operating a continuous process of self-assessment. It is a method of experimenting with a 'learning community' that requires interactive participation of the stakeholders engaged in an educational process [Livraghi, 2004].

2 Postgraduate education in support of change: personalization and interaction. The case of the II grade Master in Management of Advanced Education at Parma University

The conventional conception of education cannot be considered acceptable today in an economy of knowledge. It is no longer sufficient to talk of investing in knowledge acquired at clearly defined places and at set times [Costa and Rullani, 2000]. This is because education must respond to the continuous demand that emerges from having to cope with ever changing situations needing immediate but hitherto untested solutions. Education must therefore be continuous and personalized. In this context technologies represent an important tool for facilitating activities featuring information, discussion, collaboration and sharing.

The new University, based essentially on self-government is experimenting with a variety of educational initiatives on the net, initiatives that go from putting online material used by teachers during the conventional type of lessons in the classroom to realization of what may be termed a 'virtual' university.

We are here proposing to analyse the case of postgraduate education (the II grade Master in Management of Advanced Education at the University of Parma[1]) as an example of how users interact in a managerial model [Livraghi, 2004], a model that aims at maximum quality. This Master course lies somewhere between a university educational process of the conventional kind and that provided by a virtual university. It is based on realization of, and support from, an online learning community that came into being in response to needs expressed by a number of stakeholders interested in education.

The main purpose of this Master course is to spread the culture of knowledge-sharing, to fulfil the needs of a complex society undergoing change, supplying tools to the users and developing the potential of those transversal capabilities (not always possible with conventional education programs) needed for achieving a level of professionalism that plans to handle technological, organizational and market changes within the framework of what is known as knowledge and a creative society.

The figure that this course of education intends to create is that of *knowledge workers / knowledgė manager* able to realise the needs of stakeholders and suggest efficacious solutions suitable for application in a context both new and complex. The *knowledge workers / knowledge manager* operates with professionalism and creativity, linking up a complex network of users.

A type of education adequate for dealing with these macro-objectives must respond to the needs of users, namely:

- space-time requirements: the student attending the self-governing University is a user who chooses a course of education according to the quality of teaching available, not according to geographical vicinity. Users may be living at a distance, inside or outside the country, and be able to benefit from lessons and study material at different times;
- the cognitive and professional needs of an active and participative user who decides to extend his knowledge and skills, his ability to live and

1 The II grade Master in Management of Advanced Education was planned and carried out by Prof. Renata Livraghi of Parma University (<http://www.unipr.it>) in collaboration with CRUI, the Conferenza dei Rettori delle Università Italiane (<http://www.fondazionecrui.it>), the Consorzio Symposium di Bologna (<http://www.consorzio-symposium.it>) in agreement with the Ente di Formazione Sinform di Bologna (<http://www.sinform.it>) and co-financed by the European Social Fund, the Emilia Romagna Region and the Minister of Labour.

work, to fill a positive role in a complex and continuously changing world.

Technologies that have always made possible the removal of space-time barriers, today represent a real chance for Universities to guarantee access to continuous learning, including non-attending and worker-students, by benefiting from available teaching and services that would otherwise be denied them.

3 The context in which learning takes place

The paradigm adopted for learning is social constructivism and the educational strategy adopted is the blended or mixed form consisting of three stages:

- lessons actually held in the classroom;
- individual learning where the user relates to supporting material;
- work online that develops in learning communities for teaching purposes (virtual class).

Classroom and remote activity are spaced out in cycles: each module includes intermediate lessons and a final lesson in which one or more teachers of the module discuss the work done with the class and with the individual learning communities. The purpose of this cyclic arrangement is to:

- enable the teacher to check, to study in greater depth, and to direct learning during the intermediate remote stage, to make a critical review of the ultimate work done by the groups at a lesson in the classroom where the subjects dealt with can be discussed in a peer-to-peer relationship, and a synthesis made of the module, providing the students with an opportunity for further thought and analysis;
- enable students to attend a lesson in class, interacting with the teacher, studying the subject more fully, discussing with members of their own group and with the teacher, presenting in class the work done by the group, discussing it with the teacher and other students.

Students are organized in learning communities in order to lessen an individual's sense of isolation, but even more important to enhance the value of previously acquired knowledge to facilitate progress made by group as a whole [Harasim, 1990]. An approach of this kind is particularly suited to education of adults whose personal and professional experiences can make a useful contribution to the educational process.

The educational course comprises different ways of using the net to support learning processes according to the established program of studies and to the level of education acquired. In particular the forms of learning may be:

- independent, where the student gains access through the net to unstructured materials made available, as if from a library;
- collaborative, (more or less assisted), where the student uses materials and works towards a definite objective aided by teachers, experts and tutors;
- reciprocal [Trentin, 2004] where the student is part of a community of practices in which teachers and experts can also participate, where peer-to-peer learning processes are favoured, based on shared experiences, on knowledge and on the best practices to promote collective growth of the whole group.

We now propose to consider in detail some aspects that have proved most innovative and effective in the pursuit of educational objectives.

3.1 The platform

To ensure that a blended course functions properly great attention must be paid to the changes which the real community undergoes at the time when it becomes virtual, meaning by this, attention to planning and development of the virtual environment that will host the life of the community. In this sense the platform must be flexible and make possible 'personalization' and adaptation of the learning environment to facilitate interpersonal communication, interaction, and sharing of information.

In view of the type of user (adults, workers for the most part, situated all over the country) asynchronous communication has been preferred as, in this way, the user does not have to be online in order to participate in communication, making available to the virtual class a few simple means for collecting information and for interaction / collaboration (Trentin, 2004). These facilities are not directly related to activities of interaction and of collaborative work, but assist and help to make them more effective:

- *a space for profiles* constituting a collection of presentations and photos of the members of the virtual class, with a brief informal description (*résumé*) of each person who can then explain his own field of expertise and interests to the others;
- *a study plan area* containing a collection of indications about content and objectives of the modules constituting the educational process;
- *an area for deadlines and dates* for handing in work done at home and in class in which the tutor, in agreement with teachers and experts, inserts deadlines, details about execution of deliveries, rules governing community life.

The functional aspects of interaction / collaboration are those in which the so-called social presence[2] [Short *et al.*, 1976] of the group, necessary both for successful dialogic interaction and for collaborative work, is realized:

- *module area*: each module or group of modules has a space for discussion and making comparisons, to improve organization of communication and avoid dispersal of information and of materials concerning the themes under review;
- *the individual communities of learning and of practices*: in this post-graduate education course, considering the type of adult users and of their ultimate objectives, each leaning community is allowed to decide freely which tools to use for handling communication most effectively in support of the group's activity. This assumes that the users will have practical experience of the tools available and will behave in a responsible manner in negotiating a space for each member of the group. To solve the matter what is usually done is to open a private Newsgroup devoted to specific subjects, with an area for sharing material;
- *a coffee break* in which to discuss subjects not strictly related to those pertaining to the educational process;
- *a bibliography space* for putting together not only the bibliography recommended by the teachers, but also all the texts and links identified and

2 Taken as meaning the possibility of joining in a conversation among individuals and also handling objects (work in class and at home, sharing documents, producing artefacts, etc.).

posted up, with a short description of the content of each one to facilitate the work of anyone wanting access to it.

3.2 Personalizing the education course

One answer to the demand for continuous learning in the society of knowledge is to provide personalized education. Apart from personalizing the process in terms of spaces devoted to information-sharing, times and modes of interaction within the learning community, further consideration is being given to planning teaching activity with inclusion of a number of personalized courses, a result of the first two months of continuous interaction between class and tutor (*scaffolding*).

This has brought to light the fact that a fair number of the group in the class have specific needs. The planners decided to meet these wishes by working out together with the students several personalized courses and making available a space for setting up a Community of Practices on a theme of common interest to many students and teachers, who could create their own organization, interacting on equal terms, the community to be kept alive after the end of the education course. In this way the students feel assisted, stimulated and oriented, while responsibility is focused on users [Calvani and Varisco, 1995] so guiding them towards independence.

3.3 The role of the Tutor

In a net-learning process, the tutor assumes a role of *Community Manager* [Salis *et al.*, 2002]. He represents the community's chief interlocutor, carefully noting the requirements of both teachers and students, facilitating communication, checking on individual needs and interests, supervising motivation and involvement of members of the community so that teaching shall be organized to the general satisfaction, securing the greatest possible advantage from a relationship based on collaboration. His role is less evident than that of the teacher or expert; he remains in the background while carrying out his job of identifying and harmonizing all the activities needed for a well-run community.

His task is to lead the community in creating its own Community Memory [Trentin, 2004]. The Community Memory is a sort of sieve for the large amount of information present in different forms (digital repositories, memories and knowledge remaining in people's minds, etc.) and the daily activities of members of the group. Just as the archives of a bookshop enable any book to be found at any time, so also the Community Memory possesses its own structure and exercises control over all matters relating to development of the community's work. It is up to the community, however, to decide the basis for keeping its records, for selecting and indexing, assessing and interpreting its content as community needs dictate, as well as learning to negotiate how information and knowledge shall be handled. The handling of information is an indispensable requisite for the building and management of knowledge. The concept of Community Memory as a repository, however, is limitative: students must not see it as a container in which to insert printed or electronic material on which they can draw.

At this point three important considerations arise. The first, that quality is not only related to the material filed but is also found in the comments posted up, in the caption, in the remarks made (for instance: this document is most useful for an understanding of the XX / YY law) or in the thoughts or comparisons that the community faces. This is why, for each module dealt with by the tutor, by the person responsible for it and by the students, there must be a moment for making a synthesis of the key concepts, of the ideas that have emerged, of the difficulties that have arisen, etc., so that some trace may be preserved and improvements made in continuing the education course.

The second consideration concerns the effects of viewing the Memory as a store on which to draw without any real and spontaneous involvement of the individual as a member of the community: the wealth of material stored therein diminishes until it is obsolete and superfluous, becoming static because it has not been implemented, used and consumed.

The third consideration emphasises how essential is the role of the tutor as an initial stimulus, as *scaffolding*, but also how in time the community itself must implement and organize its own Community Memory to ensure its own survival. Exponential growth of information, in relation to the complexity of teaching aims, will also lead the community to review and rethink the criteria it is applying to organization, cataloguing and updating information.

3.4 The roles of teacher and student

The role of the teacher, but above all that of the student, changes. Each participant becomes a negotiator of meanings [Paloff and Pratt, 1999] since knowledge is built and rebuilt dialectically and constitutes the primary aim of teaching. The student is asked to make a critical examination of himself, of his style of learning and mode of interaction. This helps him to learn how to learn, enabling him to face and deal with the sudden changes that take place in the society of knowledge.

In the environment in which we work, featuring continuous changes, considerable importance is attached to a person's capacity for making relationships: capacity for communicating, for working in a group, for problem solving, for accepting new ideas and prospects.

At the university a student takes part in lessons, studies the set books and prepares for examinations. He studies alone and often uncritically. In an approach to study marked by collaboration, the teaching / learning relationship changes as does that between student and teacher:

- in accordance with times and procedures laid down at the start of any teaching work, a student can decide how and when to tackle his study material, form relationships with teachers and with the members of his own learning community. The process is one of negotiation with other members and with himself: with the others when, having established aims and timing, group decisions are taken on how best to achieve the objectives set, and with himself in discovering what value can be attached to his own style of learning and how his personal expertise and knowledge can be increased;
- teachers and experts are required to check on what is to be taught in class through active and/or passive participation in the virtual learning commu-

nity, giving help with study problems, stimulating thought and promoting reciprocal critical comparison.

The added and strategic values of transversal capabilities depend on this capacity for negotiation.

4 Conclusions

The technologies and the learning communities now in use are appearing ever more clearly as innovative tools for education in the creation of societies based on knowledge, and in responding to the new requirements of users of educational services in conditions of lifelong learning and in the 'reform' of universities.

In these situations *the student is a user who differs* from the recipient of the conventional type of formal education. Here the user is an active person who interacts and shares expertise, knowledge, solutions to problems, ideas. He is not only the user in the sense of utilizer / beneficiary, but is also a contributor. He not merely collects and uses material, but becomes an active member of a more or less extended community that actively participates in structuring, organizing and implementing material, one who is aware of his responsibilities as part of a network that can only grow if each member makes a spontaneous contribution. An active user is a valuable resource because he knows how to find solutions to problems that are unstructured or have never been faced before.

Efficient and efficacious education is the result of *joint, continuous and personalized planning* that, among the stakeholders interested in the education course (students, universities, manufacturers), has promoted communication, comparisons, interaction, so that their needs, objectives and expectations are clearly expressed. This extensive work of planning and re-planning offers users an active role, not merely as beneficiaries of pre-packaged courses; it offers them courses that are flexible in terms of content and modularity. The students of the Master course that we have analysed have in fact negotiated, together with the Executive Committee, ways and means for extending it to include two personalized courses and activation of a Community of Practices.

This project has proved a highly positive experience for the *University* that opens outward, making contact with interested stakeholders. A new and innovative experience: new in method, placing the roles of teacher and student on a level different from the one conventionally attributed to them; innovative in the use of technologies now spreading inside universities but still largely unknown; a University therefore that expresses its intention to change its methods, to become innovative and open its doors to society.

The ultimate aim behind the decision to use the *learning community* as a didactic tool is in fact that of creating a network of people and professional competences, a community of practices of experts in educational systems able to maintain contact with, and give support to its members over facing changes at their present work places or where they will be working in the future. The method and the teaching model are based essentially on the use of technologies and on interaction (peer-to-peer, class and teacher / expert / tutor), on exchange, negotiation, on creation of 'new knowledge', generated by the heterogeneous nature of the human capital possessed, on accumulated experiences, on skills, abilities, on the interests that led a person to take an active part in the education course. Comparison, debate, discussion among equals, among colleagues and among experts are all made possible by correct communication and by a form of learning that requires personal thought, by joint activity with others, creating new knowledge and social capital. Teachers and experts too experiment with a process of learning together with the students; not only therefore do they pass on knowledge accumulated over time, as happens in conventionally organized education, but also take part in a process of learning and social interaction generating in this way continuous accumulation of individual and collective human capital.

Bibliography

L. Bannon and K. Kutti, 'Shifting perspectives on organizational memory: from storage to active remembering', in *Hawaiian International Conference on System Sciences (HICCS 29)*, IEEE Computer Press, 1996.

D. Biolghini, *Comunità in rete e net learning*, Etas, Milano 2000.

D. Biolghini and M. Cenarle (eds.), *Net learning: imparare insieme attraverso la rete*, Etas/RCS, Milano 2000.

J. S. Bruner, *La ricerca del significato. Per una psicologia culturale*, Bollati-Boringhieri, Torino 1992.

A. Calvani, *Manuale di tecnologie dell'educazione*, ETS, Pisa 1995.

A. Calvani and B. M. Varisco (eds.), *Costruire decostruire significati*, Cleup, Padova 1995.

A. Calvani and M. Rotta, *Comunicazione e apprendimento in Internet: didattica costruttivistica in rete*, Erikson, Trento 1999.

A. Calvani and M. Rotta, *Fare formazione in Internet*, Erikson, Trento 2000.

A. Carrera, *E-learning dal paradigma trasmissivo a quello interattivo*, Sdilab, Venezia 2001.

M. Costa, 'Le comunità di pratica come leva per la formazione', in *Quaderni SSIS-ONLINE*, n. 3, 1999, <http://www.univirtual.it/ssis/quaderni/ssis03.pdf>.

G. Costa and E. Rullani (eds.), *Il maestro e la rete: formazione continua e reti multimediali*, Etas, Milano 2000.

P. F. Drucker, *La società post-capitalista*, Sperling & Kupfer, Milano 1994.

D. Foray, 'Una nuova capability organizzativa: il *knowledge management.* Lezioni dall'indagine OCSE', in L. Frey, R. Livraghi and G. Pappadà (eds.), *Lavoro a elevato livello qualitativo e sistemi economici/sociali fondati sulla conoscenza*, Franco Angeli, Milano 2003.

L. M. Harasim, *Online education: perspectives on a new environment*, Praeger, New York 1990.

P. Lévy, *L'intelligenza collettiva. Per un'antropologia del cyberspazio*, Feltrinelli, Milano 1996.

R. Livraghi, 'Economia della conoscenza e capitale sociale', in L. Frey, R. Livraghi and G. Pappadà (eds.), *Lavoro a elevato livello qualitativo e sistemi economici/sociali fondati sulla conoscenza*, Franco Angeli, Milano 2003.

R. Livraghi, 'The user of educational services', *in another part of this book.*

I. Nonaka and N. Konno, 'The Concept of "Ba"; Building a Foundation for Knowledge Creation', in J. W. Cortada and J. A. Woods (eds.), *The Knowledge Management Yearbook 1999-2000*, Butterworth-Heinemann, Boston 1999.

R. M. Paloff and K. Pratt (eds.), *Building learning communities in cyberspace: effective strategies for the online classroom*, Jossey-Bass Publishers, San Francisco 1999.

M. Polanyi, *The tacit dimension*, University of Chicago Press, Chicago 1975.

J. Short, E. Williams and B. Christie (eds.), *The social psychology of telecommunications*, John Wiley & Sons, London 1976.

H. Takeuchi, *Beyond knowledge management: lesson from Japan 1998*, <http://www.sveibj.com/articles/lessonsjapan.htm>.

G. Trentin, *Apprendimento in rete e condivisione delle conoscenze*, Franco Angeli, Milano 2004.

B. M. Varisco, 'Paradigmi psicologici e pratiche didattiche con il computer', in V. Minoro, G. Olimpo and D. Persico (eds.), *Argomenti di tecnologie didattiche*, Edizioni Menabò, Ortona 1998.

B. M. Varisco, *Metodi e tecniche della valutazione*, Germini e Associati, Milano 2000.

L. S. Vygotsky, *Pensiero e Linguaggio*, Giunti-Barbera, Firenze 1965.

L. S. Vygotsky, *Mind in society*, Harvard University Press, Cambridge MA 1978.

E. Wenger, *Artificial Intelligence and Tutoring Systems: Computational and Cognitive approaches to the communication of knowledge*, Morgan Kaufman Ed., Los Altos CA 1987.

E. Wenger, *Communities of practices, learning, meaning and identity*, Cambridge University Press, London 1998.

E. Wenger, *Supporting communities of practices: a survey of communities-oriented technologies*, 2001, <http://www.ewenger.com/tech>.

Barbara Rondelli, Dip. di Studi Politici e Sociali, Università di Parma

Massimo Russo

Users, new Italian subjects

Abstract:

The Italian word 'utente' *is translated by the English noun* user. *What does the noun user make us think of? When do we use the word user? Who does identify himself/herself as a user? The word user represents a repetitive and complex experience and the concept shows the changes in the Italian society. Who is the Italian user and what are his/her characteristics? What are we talking about when we talk about users? What is in common between users in a collective and an individual sphere? Today users are those people who make use of a service; they are identified and identify themselves with all the services the community developed thanks to the application of technology.*

Users want to be considered legal subjects with their own rights and obligations besides being citizens, customers and consumers; simple users. The user represents the changes in the society and responds to technological innovations. Users should be competent in deciding and operating. The strengthen of users is given by their ability to use adequate technology; this is the reason why users need to acquire confidence employing technology and turn it into an interactive transmission channel to express their needs.

Keywords:
user, technology, complexity, challenge of users

1 Introduction

The Italian word '*utente*' is translated by the English noun *user*. It is applied to several contexts, in fact, it has replaced words as customer, citizen or resident and consumer. In the beginning this noun was employed when talking about Information Technology and its applications to public services, but it has entered our vocabulary.

As a specific term it qualifies and identifies as *users* individuals who are able to combine both scientific knowledge and technical skills. At first the concept appeared in a technology related context, but it has been extended to other ones over time. It clarifies the process of development based on new technology throughout a constant definition and redefinition of who the actual users are. Both simplicity and complexity come across in the idea of user and they are not in conflict or contradiction.

Those people who have the status of *user* should be quick in becoming familiar with the rapid technology changes and are considered new 'subjects' that open the doors to new fields of study. This means that the noun conveys only one meaning, but it is used referring to several and different contexts and it appears both as ambiguous and polysemic. What does the noun user make us think of? When do we use the word user? What are its features? How do we qualify it? Who and what does it represent? Who does identify himself/herself as a user? Commonly the traits of a user are recognized in a titled individual who interacts with the general public and with all those elements that pertain to the sphere of what is considered private. Users use services developed from the application of updated technology and the growth of the service industry, considered technologically advanced, led to the accomplishment of users in a society where the demand of flexible and temporary labour is increasing. Users are technologically skilled because they belong to a community which is based on the necessity of referring to competent people. Nowadays, thanks to Information Technology, it is possible to reproduce and simulate reality, but we appeal to refined models that are intended for the future. Knowledge acquired and transmitted by new technologies widens people's view and it allows them to develop personal and individual learning according to the person's interests. The community of technology that is a sort of a knowledge community, recognizes users as active 'subjects' with their own requests that are seen as 'functional' to the community itself. The concept of user questions the traditional one based on the dynamic market-customer considered that users belong to the global world where everyone is potentially interconnected with each other and it aims to go beyond the idea of user as citizen or resident. Because of the latter a need to rethink the conception of 'fruition' and of the one of citizen-resident as the actual person who uses is perceived. The relationship man-technology is meant collectively, but it is also applied to individuals. In the show business the user is the subject we refer to when planning new ways to entertain he or she becomes a spectator and considers vision as representative of the mass. The user also gives a certain value to technological relationships and believes that this interaction attributes responsibilities.

Information Technology has given individuals the tools to extend their domain on reality because it has become available to everyone and has been introduced in every day's life. People have always been able to overcome their limits and grow to be more powerful; they have built more sophisticated instruments so that they could reach their goals limiting and containing natural efforts, waste of energy and fatigue. The history of mankind development is also made of technology; science progress and technology applications change the quality of living and provide opportunities for the future creating new expectations. Technological innovations accelerate chances and abilities to expand and change reality starting from daily life. An individual called technological individual is someone who uses and applies technology and not a passive spectator. During the 1900s mass culture had become popular, but it has been replaced by digital individuals who do the planning in an ideal and virtual context. They interact through technology. This concept is related to what's considered public, but it also applies to the private sphere.

Social science researchers often use new words to distinguish abstractions that come from social behaviours and are different from those who belong to the idea of profane.

The language of technology uses metaphors that belong to every day's life such as interactive flexibility, being able to move and have a sense of direction. This is why we use the expression 'to surf the internet'. Information Technology allows people to enhance levels of communication and information so that different subjects interact, but they are asked to take responsibility for their actions. It elaborates new settings where knowledge can take place and it widens and questions the space and time components of the human journey.

The discovery of a new and effective noun is always perceived as a conquest. The word user represents a repetitive and complex experience and the concept shows the changes in the Italian society. Users make us think of the meaning of the word '*use*' which draws our attention directly to technology and machines. The noun is most likely used when referring to public or private services and subjects offered to them.

It's not possible to give an exhaustive and clear definition of the word user. In a society full of risks, where technology investments take place on a regular bases, users express and represent the contradiction of present times. Who is the Italian user and what are his/her characteristics? What are we talking about when we talk about users? What is in common between users in a collective and an individual sphere? If the Italian word *utente* translated by the English noun user comes from the Italian '*utilizzatore*', what's the boarding line which makes the term user preferred to the other one?

Users reflect today's fragile identity of people and the need to argue traditional competences and rules acquired over time.

They, users, develop new ways of learning and as active subjects with their own rights ask for new norms to regulate interactions, exchanges and relationships. In the era of globalization they manifest new requests and each person becomes a user throughout technology interacting with machines. Today users are those people who make use of a service; they are identified and identify themselves with all the services the community developed thanks to the application of technology.

Users want to be considered legal subjects with their own rights and obligations besides being citizens, customers and consumers; simple users. The user represents the changes in the society and responds to technological innovations. When he or she is confronted with technology they claim their independence as subjects. The identity of a user is mobile and different according to the context since they are flexible and adaptable, but also ambiguous. The user-subject acts out according to his or her needs that he or she tries to satisfy rationally. Users should be competent in deciding and operating. A person is the subject and object user, an actor who interacts and communicates with a system whose efficiency and effectiveness depends on the level of interaction. As a prismatic mediator a user collects fragments of reality and tries to reorganize it. In the Italian context they embody several subjects and figures as citizens, customers and consumers and represent a whole category of problematic pragmatism and function as if an impelling problem had to be solved right away. He or she establishes a connection with politics and gives politics a concrete and not an ideal meaning.

The strengthen of users is given by their ability to use adequate technology; this is the reason why users need to acquire confidence employing technology and turn it into an interactive transmission channel to express their needs.

The concept of user as a citizen, a customer and a consumer applies to the public sector intended as the industry that provides services. Users can also be associated to the idea of voters, but those that belong to the category of voters who go beyond the ideology. They are aware that technological interconnections link different parts of the world. Transparency to acknowledge international rights is perceived as a necessity and the user shows his or her ability in developing relationships based on other people's competences as long as they are provable and verifiable qualities in front of the general public. In the complexity community, user is a synonymous of client, customer and citizen and it designates an animated subject. The political user, who goes beyond the ideology, looks at politics as being a public service and it does not express a typical sense of belonging to something. Politics is perceived as a place where personal interests merge and the user comes into play when politics turns into a show since the market looks for consent. Users appear as active subjects whose strengthen is based on their knowledge and on their resources. Political users force the traditional form and language of political interaction.

Pragmatism is the ideology of users and it supports them is their requests. Politics instead is just an active action taken in response to problems; they set ideal questions looking for concrete answers and do not like any political influence. New technology accentuates the mass character of politics which is presented more as a spectacular event, as entertainment. Politicians stand out for the projection of their figure and must enhance people's imagination. Aware of the power of technology users rely on it expressing transformations and complexities of contemporary society while looking for new trends.

2 The challenge of users

Users celebrate pragmatism by contract and their creed is on 'doing' or even better on 'giving' and use the indicative mood (grammatically speaking) when talking about something to happen in a near future. The ideology is replaced by quality of living taking into account individual parameters and starting from small and common issues. Users are conscious of their skills and are willing to make them available to public operations. Politics is considered a public service that represents contemporary society and politicians are asked to satisfy needs and interests of individuals as a reflection of the community they belong to. The requests we are talking about are on the administrative and territorial level and are related to the efficiency of institutions. Politicians see users as parts of a new way to organize and take part to operations that threatens traditional ways of being invested with a power and of empowerment. Users are aware of the effective power of knowledge and technological competences because the use of them allows control checks to be sure to have the necessary security and protection. The application of new technologies to communication permits users to be flexible' autonomous and to be able to exert their functions; they also include subjects that are usually not competent, who become noticeable all of a sudden and claim their rights. Throughout new technologies users can communicate and be informed right away overcoming space and time limits. Users are global citizens and face a lot of changes; they are aware of their competences and means; they are seen as the perfect example, the prototype, of a technologically advanced society that unites different people, that makes us think of services, applied technology and the use individuals make of it.

3 Users and complexity

Complexity has mined the bases of deterministic models. It's necessary to analyze and interpret new contexts and paying attention to the concept of user will help understand and grasp problems and dynamics of ongoing changes. The image of users is made of different and heterogeneous elements according to the different contexts where the actor, i.e. the subject, operates. They are a product of social change and express adaptability and the ability to interact not only in theory, but also practically; they gather different cultural, objective goals and neutral models; they question the inadequacy of the ways of representing the society. Their strengthen is given by all the experiences they have had and interiorized to achieve more freedom and independence. Innovation creates new forms and opportunities to what is actually possible to experience. Users become builders who put in order and give stability to the present; they are requested to have the knowledge of the operating system to be pragmatic, but they should be able to produce new legitimating values. The use of the word user correspond to the necessity to recognize the subject who interacts and uses services. This noun is becoming more and more popular and it is also captivating because it depersonalizes and dignifies the subject conferring him or her a role in society and a value for being competent and wanting this competency.

The use of a service connected to technology goes through wires and then set back on-line. We find users in the public space asking for a public or private service for individuals. The cultural context where users have become common is rapidly evolving and it shows how science and applied technology are gaining popularity. If we were to find an analogy between the two Italian nouns '*utilizzatore*' and '*utente*', we do not know what it would be. What are the differences and similarities? Users perceive the need to conform language to technological development and the ability to make and support requests. Being connected to a service and being able to use it are considered distinctive signs. Users must be interactive and their presence in space and places is widening. The individual-user represents a civil society that is disembodying and strengthening the process of individualization recognizing the importance of the single. The freedom of users comes from their independence from both as someone who makes use of a service and in the ability to choose the service they want. Users ask for and acquire responsibilities to legitimate the supply of a service; they hide behind an anonymous pin number and a password; and throughout traditional space and time dimensions. Applied technology allow users to heighten in self-reliance. Cars have shortened long distances, but today's expansion is based on Information Technology.

Citizens become users when they move around in urban spaces. New telecommunication technologies make a difference, but they also raise new problems about relationships among people. Information Technology eases bureaucracy and makes communication and information stronger, but it highlights the isolation of people in public spaces. Users have lots of competences which distinguish and qualify them; they have a strict relationship with technology applied to machines and with services that are directly offered to the public. The noun user emphasizes the feature of a new figure that incorporates elements and typical components of the contemporary society showing what is changing.

The language of users comes from the standardized language of technology and the technology community perceives them as skilled and competent people that are referred to when it's necessary to solve problems. Users replace customers on a commercial level going beyond the right of citizenship when they claim a service and manufacturers tend to refer to them as users other than customers. Such producers pay attention to the ability in transferring technology competences using communication as a source and channel to transfer knowledge. Customers rely on warranties and protection which means they rely on contracts, while users are identified by numbers and cannot count on individuality and peculiarity anymore.

Users represent an agent whose tasks are to agree, to incorporate and to understand the rules that need to be subscribed just like contracts. The society of users is a society of services with a strong interaction between public and private. They are recipients who give consent and subscribe. Some of the needs of users are private, but they operate in public and are united by common deals and interests with other individuals who operate in the same context. Users are those people who do not have the status of citizen, but cannot be denied a service as individuals. The activation of a service refers to potential and virtual users. As soon as users use a public service they require efficiency and functionality; they are asked to be flexible and adaptable, but they need to acquire competence and rules to be able to actively make proposals. Users as recipients of a service have to take responsibilities to produce and consume, but also to be interactive assistants and spectators. Their ways of communicating are based on codes, signs and knowledge so that they can express their potential to the full. Those people who cannot claim citizenship are aware of a sort of discrimination towards them and they find themselves in a grey area full of ambiguity because they are considered differently and left out.

We have here tried to give a broader definition to the concept of user which has become into use in the Italian language. This neologism replaces other words and becomes synonymous of customer, citizen and consumer and it's also used to identify the community of services and risks where Information technology skills, responsibility and security are required. Users are a new figure that incorporate many actors and factors and expresses changes in contemporary society.

Acknowledgment

The author gratefully acknowledges the help of Doctor Elisabetta Zazzeroni in translating from Italian to English. All errors, inaccuracies or misunderstandings remain, however, the responsibility of the author.

Bibliography

Z. Bauman, *Dentro la globalizzazione. Le conseguenze sulle persone*, Laterza, Roma-Bari 2000.

Z. Bauman, *Il disagio della postmodernità*, Bruno Mondadori, Milano 2002.

U. Beck, *Che cos'è la globalizzazione. Rischi e prospettive della società planetaria*, Carocci, Roma 1999.

E. Boncinelli, *Io sono tu sei. L'identità e la differenza negli uomini e in natura*, Mondadori, Milano 2000.

L. L. Cavalli Sfoza, *L'evoluzione della cultura*, Codice Edizioni, Torino 2004.

M. Gottdiener, *Life in the air*, Rowman & Littlefield Publishers, Inc. Lanham MD 2001.

G. Lanzavecchia and M. Negrotti, *La scienza e la parola. Problemi e aspetti della comunicazione scientifica*, Libri Scheiwiller, Milano 2002.

G. O. Longo, *Homo technologicus*, Meltemi, Roma 2002.

S. Zambrini, *La città in concerto*, Auditorium, Milano 2004.

Massimo Russo, LCA – Laboratory for the Culture of the Artificial, University of Urbino

Fumihiko Satofuka

Editorial. From the Artificial to the *Kyosei*. In the pursuit of the Japanese technological tradition

Abstract:

The European word 'Artificial' has been discussed by famous intellectuals in various fields and by the public in contemporary Europa since the end of the 1960's from the view points of the European thoughts for technological enlightenment. It is generally well known that the 'Artificial' is a word used originally in modern Europa after scientific revolution. However, in the intellectual tradition of European environmentalism the 'Artificial' meant the 'Alternative way of thinking for technological development' in the 1970's. And also in the 1980's it meant the 'Anthropocentric way of thinking for technological development'.

With the preceding theoretical trends the new wave of thinking was taken for the sustainability in the 1990's. It meant the 'Sustainable way of thinking for technological development' In the mean time, in the history of the Japanese thoughts for the technological developments the Japanese word 'Kyosei' (or Symbiosis) has been discussed in the various fields since 1970's. In this special issue on the 'Kyosei' our contributors also discuss the several kinds of view points. For example, Prof. Shuji Ozeki discusses the philosophical one, and Prof. Masanori Funakura discusses linguistic one, while Prof. Masayuki Horio discusses the techno-political one.

On the contrast to these points I will discuss the 'Kyosei' from the view points of the Japanese technological tradition.

Keywords
new technology culture, the *kyosei* of East and West, relation of Japanese skill and Western technology

1 Introduction

The primary concept of recent technological progress is the extent that computers have penetrated into almost all points of manufacturing, telecommunications and commodities where control is necessary. The diffusion of universal control technology can be seen for the first time.

The computer's first appearance was as calculator, and only much later became a new control technology. The computer is the first universal control device developed in the history of mankind. The influence of universal control technology is deep and wide as control is one of the fundamental elements of work; so we can characterize the penetration of the computer into control functions as a technology revolution rather than innovation. This process is changing not only the features of technology but also the face of society and continues even now. In philosophical sense it reflects the cultural inheritance of the *artificial* based on the European tradition. Since the 1970's, Japanese companies have succeeded in utilizing the revolutionary capability of computer in the development of new products.

There have been many arguments as to why Japan could realize such success. Japanese diligence, special skills, methodical nature, curiosity, talent for improvement, high level of education, and so forth have all been advanced as explanations. However, an important fact of Japanese culture has been largely ignored: Japan has a long history of isolation, and there has been a definite pattern in the manner in which she has absorbed influence from foreign civilizations. Without consideration of this factor in regard to the comparatively recent advance of Japanese technology, we will gain a superficial of understanding of the importance of the cultural inheritance of Japan. This cultural inheritance is expressed in the Japanese term *Kyosei* meaning symbiosis as the terminology of Ecology. I would like to approach the advancement of Japanese technology by focusing on this concept.

2 Japanese technology as the combination of skill and practical knowledge

The theoretical, analytical and linguistic knowledge has been important in the tradition of European fields of thought since the Greek philosophers. It is needless to say that this knowledge bore the fruits of modern scientific knowledge. In Japan, this kind of knowledge was not developed, therefore intuitive, practical knowledge was important. *Waza* was important in arts and *budo*. *Waza* is the way to play dance, to operate sword in *kendo* or Japanese fencing, to play the tea ceremony, and so forth, and integrates skill and practical knowledge. This *waza* is common in craftwork, too. In the craft, it might be better to say *takumi* rather than *waza*.

Logical concepts and scientific knowledge have linguistic and mathematical expression in the western countries. In comparison with this, practical knowledge grasped by intuition is expressed in the 'pattern'. The 'pattern' is universalized shape, a key concept in Japanese culture. It is said that the tea ceremony and traditional Japanese flower arrangement, for example, express the beauty of pattern. A Japanese sword has a special curvature, and it is said to be the reason for its excellence cutting quality and beauty. The special pattern of its curvature was selected from numberless shapes by not logical thinking but intuition, and this became the pattern of the Japanese sword. Truth and calmness of the cosmos is grasped by intuition and is expressed in the composition of the trees, stones, water and moss in Japanese gardens. Bonsai expresses harmony of nature in pattern.

Thus, in Japan, knowledge is not theoretical and analytical but practical and presented not in language but in pattern.

Michael Polanyi introduced the concept of 'tacit knowledge' for understanding. He pointed out the importance of non-analytical, integrative knowledge in addition to theoretical knowledge and understanding. In Japan, intuitive knowledge was remarkably sharpened sophisticated because of the lack of theoretical knowledge, and it was presented in the pattern.

We can characterize this kind of Japanese mental attitude to the nature as the *kyosei* (symbiosis). There are three reasons why the *kyosei* well developed in Japan:

1st historical conditions;
2nd natural environment;
3rd language.

3 Japanese technology and Western technology

Japan had no tradition of developing machinery, while western countries had depended on it in their tradition of technology. As a result, Japanese developed manual skills instead of machinery. It is natural that this skill was combined with practical knowledge and created a high level of *waza* under the influence of Taoism.

At the end of Edo period, western people who visited Japan were surprised by the high quality handicrafts they found, handicrafts whose quality was a result of *takumi*. These handicrafts had precise finish, enough solidity for practical use, functional beauty and artistic beauty as well. This was made by laminating two kinds of iron; hard steel and soft iron. This laminated edge was made easily sharper by high level of sharpening technology. The excellent cutlery is a technological base for manufacture. Perry, who was given a lacquered pen box from Edo government, said that 'Japanese people showed high level of dexterity in the practical arts, and handicraft technology was complete'. C. P. Thunberg, a Swedish medical doctor, said

> Japanese are good at applications. They have no talent of invention. Iron and copper products are better than European. Swords, clocks, glass work are excellent, too. Lacquered vessels are of a unique quality not seen in foreign countries.

A. V. Siebold recognized that 'the Japanese have a specially strong desire of knowledge, love for things'.

The Japanese tradition of *takumi* based on the *kyosei* thinking and skill and western tradition of technology based on theoretical thinking and machinery shows a sharp contrast. It is possible to show this relation in Fig. 1.

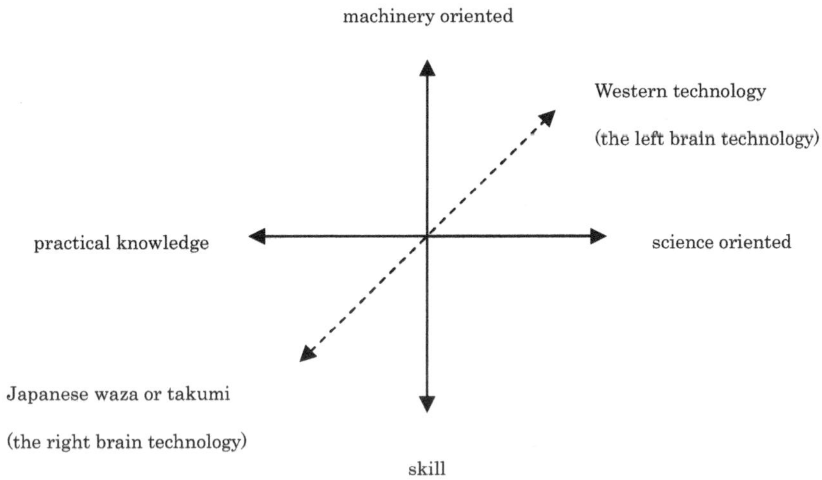

Fig. 1. Relation of Japanese *takumi* and Western technology.

Takumi is almost same as *waza* and can be expressed as right brain technology. On the other hand, Western technology could be expressed as left brain technology.

4 Dual layer structure of Japanese technology

Technology has the dual layer structure as shown in Fig. 2.

The total triangle shows the existent technology. The basic layer consists of traditional, inherent features of technology formed from historical, social and climatic conditions in a country. It plays part of the buss in a symphony. It is similar to the massive unconsciousness in psychoanalysis of the Jung school. The upper layer consists of features of technology added later. The existing technology has all this dual layer structure in itself.

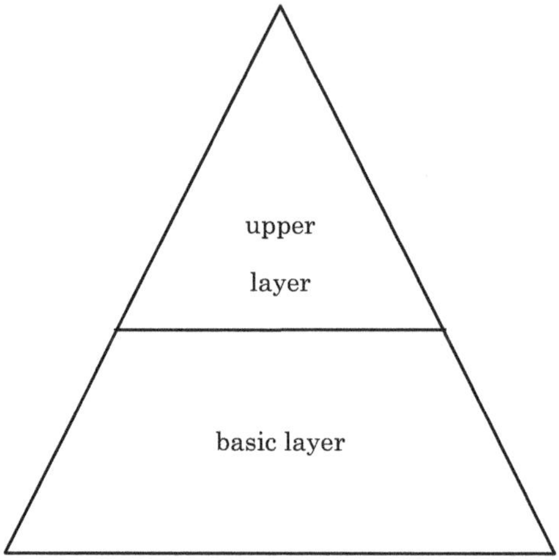

Fig. 2. Dual layer structure of technology.

Therefore, the existing technology is individual, although scientific knowledge is universal. There is no exception even in case of high-technology. It is a useful trial to see why Japan succeeded in the development of high-technology. For example, the development of N/C equipment in Japan can be explained by the dual layer structure of Japanese technology. In Japan, N/C equipment was developed by Fanuc pursued a simple and cheaper version based on open-loop control using a new stepping motor. In this case there was a big difference between the US and Japan. In the US, closed-loop control was adapted, that was more complicated both in theory and mechanism. On the other hand, Fanuc stressed practical usefulness, even if it was more simple in theory. Further, Fanuc promptly harnessed the greater capability of micro chip and could utilize the Japanese traditional craftsmanship. We can see here the *kyosei* of the upper layer (high-technology) with basic layer (craftsmanship) in Japanese technology.

5 New Technology Culture in Japan

The basic character of Japanese technology rests on the idea of '*kyosei*'. In Japan, a new culture has been created by the *kyosei* of different cultures, just as a new technology has been created by the *kyosei* of different technologies. This '*kyosei*' can be described as a basic facet of Japanese culture-creativity.

Recent Japanese technology is based on the *kyosei* of western industrial mechanical technology and Japanese craftsmanship combined with the *kyosei* of scientific knowledge and practical knowledge as well.

When this *kyosei* creates a new technology or culture, the *shu-ha-ri* (conserving, breaking, separating) principle that is inherent thought in the tea ceremony is adapted. The *shu* phase concerns an amount of straight imitation. The *ha* sees an introduction of creativity into the process. With the *ri* phase, a completely new and independent path is chosen. It could be said that Japan entered this *ri* phase during 1970s. During this decade, Japanese companies rapidly developed such new products as: N/C equipment, audio visual equipment, business machines, digital equipment such as new type of camera and watch, stepper, and so forth, relevant aspects of Japanese traditional craftsmanship are:

1. high quality raw materials;
2. high quality tools (e.g. Japanese laminated cutlery and fine sharpening technology);
3. skill and practical knowledge;
4. perfect finishing technique and attention to detail;
5. durable products combining practicality and beauty.

Artisans can easily produce items of high quality by means of highly – developed skills, particularly if good materials and tools are available. The above mentioned aspects of craftsmanship also hold for the manufacturing of products. There requirements are the same whether you are producing goods by hand or through the process of modern industry. The gist of new technology is that this culture rests on industrial products produced by the *kyosei* of modern science and industrial technology on the foundation of the traditional Japanese work ethic.

The concept of '*kyosei*' (symbiosis) is the key word for understanding the special feature of Japanese new technology.

Bibliography

M. Negrotti, *Naturoids: On The Nature of the Artificial*, World Scientific Publishing Company, New Jersey 2002.

M. Polanyi, *The Tacit Dimension*, Routledge & Kegan Paul, London 1966.

Saegusa, Hiroto, Saegusa Hiroto Chosaku-shu, Dai 10, 11, Gijutsu no Rekishi (History of technology), Chuo-Koronsha, Tokyo, 1973.

D. T. Suzuki, *Zen to Nihon bunka*, Iwanami Shoten, Tokyo 1964.

T. Watsuji, *Fudo,* Iwanami, Tokyo 1935.

Fumihiko Satofuka, Department of Ecoregion Science, Tokyo University of Agriculture and Technology

Masanori Funakura

Kyosei, or association in shared environment – from the viewpoint of ecosemiotics

Abstract:

Characteristically, the concept 'kyosei' as a translation of symbiosis is used to consider two kinds of 'environmental' issues, natural and human: humankind's 'kyosei' with nature and with each other. However, symbiosis is the fact in itself in the natural environment, while 'kyosei' becomes the fact to be explained and reconsidered when used in the human context. So this article proposes an alternative model to reconsider the concept 'kyosei' from the viewpoint of ecosemiotics, assuming semiosis running through the continuity from natural to artificial signs. Semiosis-based understanding of the 'kyosei' relationship suggests that every organism exists inseparably from the environment in its biological and semiotical process as well. Qualified by recognition of the human double role, both one part and interpreter of the natural environment, our anthropocentrism sets up the participatory association-type 'kyosei' strategy, a different model from the mutualism-type 'kyosei' one.

Keywords:
kyosei, semiosis, habit, strategy, symbiosis, association, participation, artifact

1 Introduction. An anthropocentric hypothesis

Recently in Japan, there was a sudden upsurge of public interest in 'kyosei', or symbiosis. A sense of anxiety about the devastation of the natural environment seized many people and the mass media began to spread the alarm that what the disastrous results will be is too obvious to be ignored. However, besides our 'kyosei' with nature to survive, our 'kyosei' with each other has come to be called for in the presence of unfair treatment in the social environment, that is increasing sense of unfairness caused by the aggrandized difference of wealth between the stronger and the weaker. Thus, the establishment of the 'kyosei' relationship is required to solve those two kinds of 'environmental' issues, natural and human.

Symbiosis is the fact in itself in the natural environment, while 'kyosei' in the sense of symbiosis becomes the fact to be explained and reconsidered when used in the human context. So, to propose an alternative model to reconsider the concept 'kyosei', this article starts by finding fundamental interrelations at various levels between organisms and the environment from the viewpoint of ecosemiotics, the study of the organism-environment semiotic interrelations. Assuming the continuity from natural to artificial signs, our ecosemiotic approach to semiosis or sign interpretation that differentiates each organisms' niche from other organisms' and that develops its integration of differentiation.

Probably, our ecosemiotic strategy may be open to serious criticism at once: the assumption of semiosis as an underlying phenomenon to study the organism-environment relationship is a kind of 'vitalism' based on the idea of the existence of irreducibly biological facts inexplicable in chemical and physical terms. Semiosis-based understanding of the 'kyosei' relationship, however, does not suggest the doctrine of such a spiritual autonomy of all parts of nature, but the assumption that every organism exists inseparably from its environment to the extent that it seems 'doomed' to act and react not only in its chemical and physical but also biological and semiotic processes.

There may still remain suspicion that such a hypothesis is anthropocentric, so unscientific. However, our anthropocentric view is different from such as that of Christianity, where human is not simply part of nature but made in God's image. Humankind has been seeing the natural and artificial environmens from our own point of view; indeed, we cannot grasp them independently from our 'subjective' viewpoint. Any conception, even its 'objective' data of the environment can not help being less or more anthropocentric at its base since they can be constructed only on our 'subjective' experiences. Yet, we can also adopt some holistic viewpoint, actually impossible to experience but imaginable, which has underlined modern science. Thus, our anthropocentrism is qualified by recognition of our double role: both one part and interpreter of the natural environment; or, metaphorically, one of the 'apartment dwellers' and the 'manger' entrusted the right of management of the whole apartment by nature, its 'owner' unable to speak and manage.

2 Ecosemiotics based on Peirce's triadic relation

Now, the identity of a habit depends on how it might lead us to act, not merely under such circumstances as are likely to arise, but under such as might possibly occur, no matter how improbable they may be[1].

In brief, according to Charles S. Peirce, semiotics is a science about

> anything which determines something else (its *interpretant*) to refer to an object to which itself refers (its *object*) in the same way, the *interpretant* becoming in turn a sign, and so on *ad infinitum*[2].

The triadic relation of sign (*representamen*)-object-*interpretant* is semiosis, in which everything creates in the consciousness or mind of the agent 'the equivalent or developed sign, or interpretant'[3]. A subsequent sign interprets each antecedent sign, and then the sign-object relation becomes the object of a new sign as a result of a kind of translation: a sign copied but different from the original one is produced. This is pivotal to Peirce's semiotics, who develops his general theory in a wide range including ecology as the object of its study: 'The universe is perfused with signs, even if the universe is not composed only of signs'[4].

1 C. S. Peirce, *Collected Papers. Vols. 1-6*, (edited by C. Hartshorne and P. Weiss), Vol. 5, Sec. 400, Harvard University Press, Cambridge MA, 1931-1935.
2 *Op. cit.*, Vol. 2, Sec. 527.
3 *Ibidem.*
4 *Op. cit.*, Vol. 5, Sec. 448, fn.

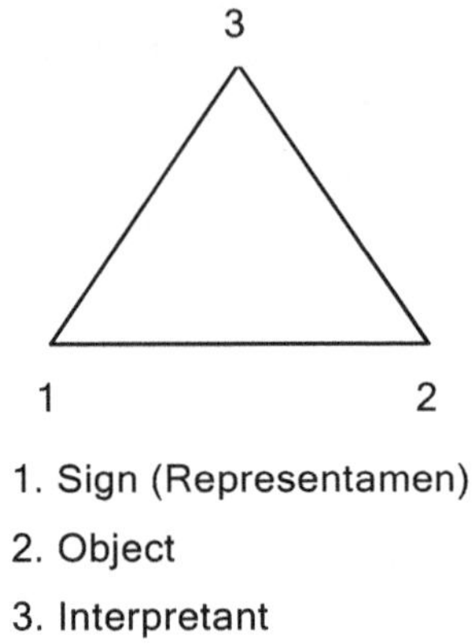

Fig. 1.1. Triadic Relation of Sign.

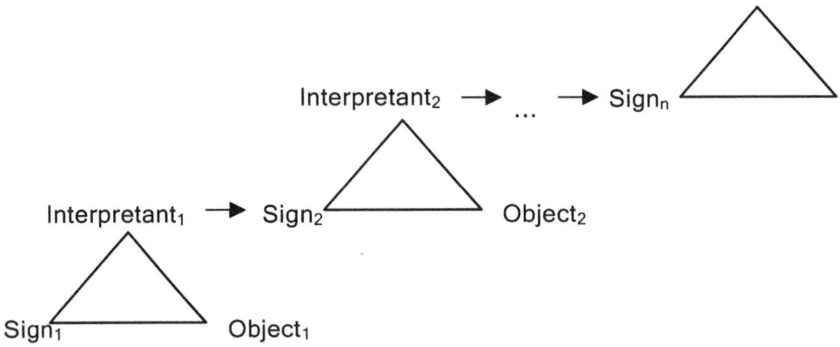

Fig. 1.2 Semiosis

Based on this general semiotics, ecosemiotics makes a fundamental hypothesis that there was the origin of a 'subject' or 'agent' able to interpret the environment of external things or conditions as giving some effect to its being or activity. In the sense that any 'subject' can transform the dyadic relation between the organism and mere fact or chance in reference to its purpose or goal, semiosis is not restricted just to the process in higher-dimension organisms. So we must take notice of the astonishingly dynamic triadic relation, or semiosis between the organism and the natural environment.

The organism maybe considered separated from its environment, but, through complicated interrelations, the system can assume almost autonomous states with regard to the environment as well as states due to environmental influences or aiming at an effect on the environment, and 'the *development of semiotic systems* (semiogenesis) originates' from 'certain inner states which are autonomous with regard to the environment, but whose interaction with the environment would be advantageous to the organism if such interactions were possible'[5]. In this sense, we must consider both the organism's practical interactions via its perceptions with its inner space as well as with the outer space or the environment as the total sum of biological and physical factors.

In this complex dynamism, each organism's or population's adaptive evolution gave rise to diversity by interpreting and modifying almost the same environmental changes according to its specific needs, perception, and ability at the moment. In the ongoing process of semiosis, however, any organism may provisionally adjust itself to the relation with its environment, and this provisional collection of interpretants may form a 'habit', or the ground of subsequent perception and interpretation. 'If signs were not habituated, living systems would very quickly degenerate into undifferentiated and plasmatic chaos. To live is to be a creature of habit'[6]. The habit is an organism's economically effective way to cope with its environment, which is acquired as a result of the organism's taking a regular tendency towards pattern and order. Not only that, but also, importantly, Peirce's use of 'habit' as 'an organizational principle that could bring an organism's models of response, or its interaction with an environment, into a unified, strategic rule of

5 U. L. Figge, 'Semiotic principles and systems: Biological foundations of semiotics', in W. Nöth, *Origins of semiosis: sign evolution in nature and culture*, Mouton de Grueter, Berlin-New York 1994, p. 27.

6 S. Jarosek, 'The Laws of association of habits', in *Semiotica*, 133-1/4, p. 83, 2001.

action' suggests 'that in many cases the notion of strategy and the notion of habit have overlapping meanings'[7].

Ecosemiotic understanding of the organism-environment relation is understanding of semiosis as a material process, where sign and 'habit' are used as material phenomena and strategies, respectively, to succeed in taking part of the eco-sytem. Because the organism cannot perceive the whole information about its environment, making a decision and action to get feed in reference to some information from the environment may lead to uncertainty, that is the risk of being eaten by predators. So to deal with uncertainty appropriately in success, it is thought necessary for the organism to find a certain constancy or regularity as safety base.

Thus the organism taking the 'habit' of constructing the properties of its ecosystem from the samples of it parts is its adaptive strategy for the environment. A series of action may lead to positive rewards such as feed, water, safe hiding-place, opposite sex, and resources useful for survival and reproduction, while another series of action may lead to negative ones such as predators or accidents. To survive in such a dynamic risky feedback system, the organism must use the already known pattern of action-reward and, if necessary, restart semiosis and explore the unknown pattern. So it is considered important to keep both patterns in balance, which is often not so easy to keep.

As the environment reveals its multiple structure to organisms, each organism reflects its properties in its structural relations to the environment. Thus, the behavior of any organism is environment-dependent from the moment it came into being as a species. So the 'habit' itself can be represented as a kind of strategy for 'kyosei', symbiosis with the environment in the sense that those diverse structural relations to the same environment may necessitate 'kyosei', or symbiosis (which is used for the time being in this paper) as a strategy for securing its niche.

7 A. Pietrinen, 'Peirce's game-theoretic ideas in logic,' in *Semiotica*, 144-1/4, 2003, pp. 36 and 45.

3 Kyosei: symbiosis or Association

Now we clarify our thinking of the concept 'kyosei' used widely in Japan today by looking at its underlying implications. Probably, 'kyosei' is used as a Japanese translation of the biological term, 'mutualism' (win-win relationship), one of three 'symbiotic' modes of living together between two species of organisms: the other two are parasitism (win-lose relationship) and commensalism (win-no win or -lose relationship). Considering that any organism or population cannot avoid influencing both other organisms and the environment by competing for resources in its own ways to survive in each niche, it is interesting indeed that the principle of competition-exclusion or the natural course of a food cycle on one hand and 'mutualism' on the other hands inhere in the natural environment. Considered thus, transferring the concept symbiosis into the human context with emphasis on just 'mutualism' may run a risk to simplify the biological fact because symbiosis co-occurs with the survival of the fittest or competition in the natural environment.

Our ecosemiotic perspective is quite different in considering that the human race together with their artifactual ecosystem constructed through semiosis is also wholly and ineluctably embedded in the natural ecosystem and that, in such a multi-interrelated system, the relationship not only between the human and the natural ecosystem but also among human members in it are too subtle and complex to be located simply in symbiosis, and 'mutualism' in particular.

Here is a very enlightening explication of 'new levels of biological organization' created by 'evolutionary changes'[8]:

- The emergence of self-replicating molecules paved the way for the evolution of self-reproducing molecular communities.
- These evolved into the ancestors of modern bacterial cells. Bacterial cells evolved into higher, nucleated cells, perhaps by forming *symbiotic communities* [emphasis mine].
- Communities of higher cells became simple multicellular organisms like sponge.

8 D. Layzer, *Cosmogenesis: The Growth of Order in the Universe*, Oxford University Press, Oxford 1990, p. 212.

> Biological organization of life is a functional hierarchy and biological evolution is a process of hierarchic construction in which new function units result from the differentiation and integration of existing function units. Differentiation and integration are closely connected[9].

Thus, every organism is 'inscribed' in the natural environment just like a pattern woven into a carpet. At the top or end of this continuum, humankind evolved into being via semiosis as differentiation-integration of signs from the environment, and artifacts embodying the results of semiosis would have compelled humankind to follow the pattern of direct or indirect relations in such a closely interconnected biosphere.

This view of hierarchic biosphere assumes that points and events nearer in space and time are more likely to be linked by interrelations, to co-participate in larger points and events, and to constitute the next larger system level. Any organism at any level is one complementary sub-system of the bio-system. In other words, the systemic world is a kind of pyramid system based on effective control of sub-systems by the system itself as the ruler. Any sub-system does less or more damages to the super-system by their behavior. Despite such feedback from sub-systems to the whole system, they are never autonomous, but restricted to a greater degree by the whole system.

However, every time new types of 'symbiosis' were born, the bio-system greatly increased its diversity. So the relationships among organisms changed from one-sided to both-sided one: for example, animals carry pollen and seeds as well as plants offer feed to animals. Thus, each sub-system in turn exerts some, even if trifling, influence on its own background and structure, so qualitative changes flow in both directions. Under the influences of such an environmental background and structure, each organism can rely on its relative stability of its shared environment, which allows both the survival of the fittest and symbiosis to be moderately in balance. If predators or parasites should eat clean their victims or hosts, they will also die away naturally. In this sense, 'kyosei' is considered a

9 *Op. cit.*, pp. 36 and 38.

natural 'habit' or contrivance to avoid 'kyoshi', or common ruin in the bio-system.

Such a view makes us reconsider that humankind cannot help being dependent on the eco-systemic network just like any other organism. But human business of life has changed this biosphere by its extensive interaction with it to the extent that will drive it to 'kyoshi', or common ruin: exploitation of the natural resources from the ecosystem, continuous from early hunting-gathering through farming to technological developments of artifacts. Thus, the natural ecosystem has been greatly altered with the result that most of the plants are artificial in the sense that human has cared for. Indeed, human dramatic development has since caused more habitat destruction with the result of impoverishment of the wildlife in many ecosystems, let alone the pollution put on the natural environment As a result of the conspicuous unsoundness of the natural ecosystem, human sustainability itself will be greatly lessened. It is only humankind, however, that can also care for or manage the whole system. Unless we recognize that biosphere can not be reset, we may not be able to escape the outcome of 'kyoshi', or common ruin.

The key to recovering the human-environment sound relationship is management and caring to keep harmony with the activities of other organisms in the scale of sustaining the bio-system as it is and securing human's interests. In the sense that not merely 'kyosei' with other organisms or nature but also 'kyosei' among members in nature must be promoted, the concept 'kyosei' requires being placed in a new context. So we should develop the active 'kyosei' relationship from this high-dimension relation available to humankind as a sufficiently advanced organic system: 'association' in which humankind permanently lives with many different organisms, always managing and caring for the system including them. In the social scale, it is a kind of colony supported by feeding the constant traffic of semiotic inputs and outputs. So our model of the active 'kyosei' or a higher level 'association' is fundamentally built with collective artifacts embodying human semiosis.

4 Kyosei from 'between' to 'among'

> [...] each human being similarly participates in an inseparable way in society and in the planet as a whole. What may be suggested further is that such participation goes on to a greater collective mind, and perhaps ultimately to some yet more comprehensive mind in principle capable of going indefinitely beyond even the human species as a whole[10].

The popular but one-sided concept of the 'between'-type 'kyosei', or 'mutualism', is unable to recognize the heterogeneous activities 'among' the natural ecosystem members. Instead, our model presents an organic conception of humankind's relation to the bio-system in that we and our artificial ecosystem can survive only in correlation with and embedded in, the bio-system whose resources are continuous with human life as the staple.

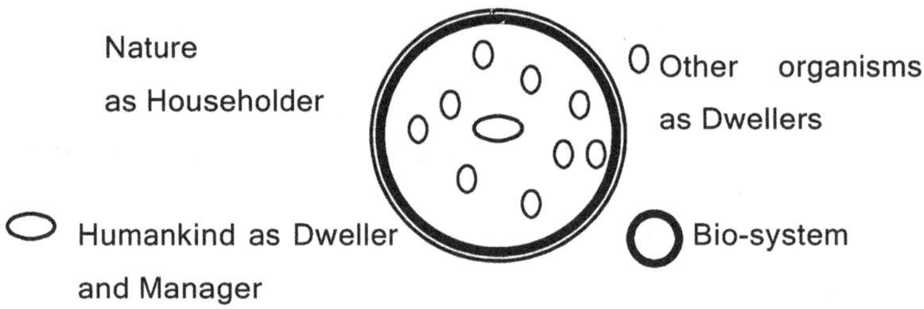

Fig. 2. 'Kyosei' among other organism in Nature

If we do not consider our life together with other organisms in the natural continuum, we can not give a new development to the concept of 'kyosei'. To develop a new 'kyosei' strategy from 'association', we must first reconsider the significance of active participation in nature. Here we examine two examples of understanding the concept 'kyosei' from the Buddhist perspective: 'so-soku and so-nyu', or the origin of all things and 'shin-nyo', or the quintessence of the universe.

10 D. Bohm, *The Essential Bohm*, (edited by L. Nichol), Routledge, London and New York 2003, p. 386.

First, a Japanese farmer, J. Miyamoto criticizes the 'city dwellers' ' attitude to 'take nature as an object for recognition and carelessly express humans' *kyosei* with nature' because 'those who live together with environmental phenomena do not use such an expression'[11]. This criticism points out that the natural environment that mattered most to 'city-dwellers' is one beyond the city limits, that is just a world that exists apart from and for the benefit of them. His criticism comes from his 'self-sustenance thought' based on the world-view preached in Kegon-kyo, or the Avatamaska Sutra. Its important idea of 'so-soku and so-nyu', or all in one, one-in-all may be summarized as follows:

- In the order of so-soku, the existence of 'Many' embodies itself in virtue of that of 'One'; the existence of 'One' is knowable in virtue of that of 'Many', so the existence of 'One' and 'Many' are closely and inseparably connected;
- In the order of so-nyu, 'One' and 'Many' are also inseparably connected because the function of 'One' influences that of the whole and the function of 'One' is knowable from that of the whole.

In other words, 'all things look woven into each other just like the meshes of a net'. The relation between 'One' and 'Many' is also explained as 'ichi-soku-ta', or 'One is Many' and 'ta-soku-ichi', or 'Many is One'. This idea has been accepted and inherited in Japanese's view of nature[12]. Explaining it metaphorically, 'If we lift up point A of a net, all other points correspond to each other infinitely and inter-relatedly around this point, and the same relation holds good both at point B and C, *ad infinitum*'.

Second, a Japanese ethicist, S. Kameyama proposes 'a fundamental framework of human-nature relationship' based on the idea of the world origin again, beyond the mechanical opposition between human-centrism and nature-centrism. His idea of 'kyosei' relationship derives from both the ideas of 'engi', or phenomenal

11 J. Miyamoto, 'Commencement of Self-support Philosophy in the Twenty-First Century: Proposal of Net-working Fifty Three Agriculture Road All Over Japan', private press, 2003, pp. 43-46.

12 S. Kamata, *The Avatamaska Philosophy*, Kodansha Gakujutsu Bunko [Scientific Book], Tokyo 2003, p. 86.

entities being interdependent on each other and of 'gu-shou', or everything being born and living together[13]. Substituting the word 'nature' with 'the natural environment', this idea seems to correspond in basis to that farmer's idea of 'kyosei' relationship with the natural phenomena.

In absolute contrast to Judeo-Christian thought fostering the aggressive, human-centered attitude toward nature, the Buddhist ideas neither claim human's superiority to nature under a dualism of human and nature nor insist that human can exploit the natural environment at will. Without being liberated from such a Christian-like one-sided anthropocentric attitude, it is doubtful that just more science and technology will get us out of the environmental crisis. What we should do about the natural environment depends on our reconsideration of human-nature relationship.

Recognizing that the worldviews of quantum physics and Eastern ideas are similar in explaining external phenomena such as interrelations in nature in particular, S. Kamata points out 'the fundamental difference' in their standpoints: theoretical physics aims at inquiring the truth of the universe while the Avatamaska Sutra aims at emancipating humankind[14]. However, a British quantum physicist has insight into human mind as well. David Bohm's 'overall approach that encompasses all aspects of objective nature and of our subjective experience'[15] will allow us to understand 'not only inanimate nature, but also living nature' from an alternative to Christian view of human-nature relationship. The key notions are the 'implicate order', where 'everything is enfolded into everything' and 'the explicate order', where 'things are unfolded in the sense

13 S. Kameyama, 'Deepening the idea of Kyosei and the significance of the Buddhist Ideas as a Frame of Reference', in S. Ozeki et al. (eds.), *Search for 'Kyosei' ideas: From the Asian Angle*, Aoki Publisher, Tokyo 2002, pp. 107-108.

14 S. Kamata, *Op. Cit.*, pp. 38-39.

15 D. Bohm and B. J. Hiley, *The Undivided Universe*, Routledge, London and New York 1993, pp. 388-389.

that each thing lies only in its own particular region of space (and time) and outside the regions belonging to other things'[16].

From the Latin root of the word 'subtle', to *sub-texere*, or 'finely woven', Bohm explains the 'implicate order' metaphorically: 'Each can "catch" a certain content of form and structure of what is "caught" in the coarser nets; they can also hold within them a further content that is implied in the latter'[17]. This metaphor of 'nets' coincides with that of the Avatamaska Sutra. Bohm himself recognizes his thought being close to Buddhism, Tibetan one in particular:

> The point about the Buddhist philosophy is that they [Krishnamurti and the Buddha] have the notion of mutually dependent origination: everything originates together, mutually dependent. I think this is very close to the implicate order, which says that everything comes out of a ground and everything is interrelated, and that underlying it there is no substance that can be defined[18].

Thus his philosophy corresponds fine to the idea of 'One-Many' interdependence in the Avatamaska Sutra.

Naturally, 'The implicate order implies mutual participation of everything with everything. No thing is complete in itself, and its full being is realized only in that participation', because 'only in the whole that the thing is complete'[19]. This participation allows each member to make some contribution to the shared system, which is too subtle to be irreducible just to a particular point of the whole effect of the whole order. Each participant depends on the interrelations with others just like one point or mesh of one whole net. Each organism among the ecosystem and each member among the socio-system can be considered not merely interacts with others in 'the implicate order', or 'One', but also takes a part of 'the explicate order' or 'Many' and adding a new order to the whole by always introducing new diversity and variation, driven by the imbalance

16 D. Bohm, *The Essential Bohm*, (edited by L. Nichol), Routledge, London and New York 2003, p. 85.

17 D. Bohm and B. J. Hiley, *Op. Cit.*, p. 177.

18 D. Bohm, *On Creativity*, (edited by L. Nocol), Routledge, London and New York 2004, pp. 132-133.

19 *Ibid.*, p. 130.

between the orders of inputs-experiences and of outputs-actions. Such dynamic interrelations among the shared space is a feature clarifying the creative activities of semiosis, each participant adding a developed interpretant to the established habitat or 'habit'.

Certainly, in the long-term evolution, humankind has changed much of the natural system into the artificial one suitable to live in by intervening or caring for it. And humankind as a technological species has accumulated many skills and artifacts to transform, utilize and exploit natural resources and forces. Unless we first recognize that every human artificial system is surrounded by not only social, political, and economical, but also biological spheres and that any action taken at least by two agents requires sharing the very space, of which each of them can take part, and then substitute the idea of manager-participant for that of conqueror-invader, however, our technology-based approach to the bio-system will fail in sustaining our living space.

5 Strategy for sustaining society in nature

> I wonder if everything forgotten was unnecessary and out of date. I am often afraid that a superstitious belief in progress mistakes for progress even what is retrogressing and that such a belief may at times drive away not only human beings but also every living thing to the point of extinction[20].

This is a protest by a Japanese folklorist, T. Miyamoto. About twenty-five years ago, he already pointed out the serious situation that old good human relationships such as mutual understanding or solidarity were being forgotten under the modernization and utilization brought about by technological progress. Although he never used the word 'kyosei', the necessity of 'kyosei' is implied in his insistence on recovering and reevaluating 'important values or meanings [...] that may be included in what many people are now

20 T. Miyamoto, *Journey of a Folklorist*, Kodanshya Gakujutsu-bunko [Scientific Book], Kodansha, Tokyo 2000, p. 235.

forgetting'[21]. So, appealing to 'kyosei' today seems to counter-prove the growing loss in the context of technological modernization of the once shared social living space. In this sense, the environmental crisis in essence is not only a technological but also social and individual problem.

What has been valued most in Japan following in the footsteps of America is growth, development, change, novelty, innovation, wealth, power and so on. 'Many people believe as if everything is progressing. Such a mistaken belief makes humankind arrogant and in turn be under an illusion that such arrogance is a privilege of civilized society'[22]. The private profit or success dominant in our modern system can lead to the violation of not only ecosystem's but also sociosystem's standards. The mistaken belief in the unlimited natural resources and the relative absence of philosophical restraints have built a faster-developing, more mobile, and more innovative society. This tendency is reflected in every aspect of our lives because our expansionary ethos has stimulated the reckless, uncontrolled growth of each member in the social ecosystem.

At the level of the social system, this popular measure of success means gaining control over a collection of inherited shared artifacts that semiosis has created. Artifacts have been designed according to interpretants believed correct and filled the 'explicate' social ecosystem, which both produce and model the 'implicate' processes as well as which connects the 'implicate' processes to each moment. Such artifacts are subsequently used to further shape materially not only the socio-system but also the natural system. It is this double-connection that always increases the complexity of our socio-natural system interrelations because the socio-system is composed of artifacts which has been significantly structured by exploiting the natural system abundantly.

In Bohm's terms, 'the question is, how do we manage this one household which is the world? But the first thing is to see that it is one. If we pretend it is many we cannot manage it'[23]. In other words,

21 *Ibidem.*
22 *Ibid.*, p. 203.
23 *Ibid.*, p.135.

nature is public property. So the problem is whether humankind has treated our habitat with philosophical restraints in acquiring natural resources from the earth. Under the circumstances, the following suggestion is understandable enough that humankind as 'the so-called "semiotic animal" ', both 'do justice to the human capacity for semioethics on a theoretical level' and 'evidence our vital, inexorable need for this semioethical capacity [...] beyond the established order, beyond the symbolic order, beyond convention and habit': the ethic that 'implies a relation of involvement and responsibility with [...] all of life generally over the entire planet'[24]. In essence this proposal of semioethics seems to be an indictment of the destructive, power-oriented uses of scientific and technological knowledge, which is due to the arrogant conception of human as wholly unique.

A new participation-based and association-style 'kyosei' strategy must be developed from the viewpoint of the qualified anthropocentrism in order for it to be really practicable, because the survival of human socio-system is the premise for conserving the natural eco-system. So Wilson's proposal of 'the necessity of a robust and richly textured anthropocentric ethic-one based on the hereditary needs of our own species', *'Biodiversity' as 'the frontier of the future'* along with environment conservation from 'selfish altruism' aiming 'to manage wildlife to the benefit of mankind'[25] is welcome indeed, as long as we can do so by managing and caring for the natural ecosystem under the agreement that the natural ecosystem is public property.

Then, how should we make the practicable frame of reference that acknowledges the inevitability of less or more competition-conflict on one hand and the prescriptive measure to encourage many members into cooperation on the other? This difficult integration of two measurers will be a key to shifting from the conquest

24 S. Petrilli, 'Semioethics, subjectivity, and communication: For the humanism of otherness', in *Semiotica*, 148-1/4, 2004, pp. 73 and 87.

25 E. O. Wilson, 'Biophilia and the Conservation Ethics', in S. R. Kellers and E. O. Wilson (eds.), *The Biophilia Hypothesis*, A Shearwater Book, New York 1993, pp. 38-39.

or invading type to the participation or sharing strategy, or the association style 'kyosei' managing biological resources of its own land.

The 'kyosei' model here affirms the need for human observation of their limits set by its ecosystem, opposed to the strategy of industrial capitalism placing the highest premium upon ingenious methods for circumventing those limits. The non-feedback conquest style system will end in 'kyoshi', or common ruin; while association style one will be able to avoide 'kyoshi', or common ruin by respecting diversity through every-level feedback to restrain human activities to tend towards the conquest type one. That is, if a system does pitifully little to encourage or reward the constraints necessary for the long-term ecological sustenance of society, it will inevitably lead to 'kyoshi', or common ruin (Fig. 3). In the face of the undeniable ecological fact of the exhaustibility of the natural ecosystem, the conquest type system seems to have already led to its bankruptcy.

In order to avoid kyoshi, or common ruin, our primary norm must include the conservation of natural ecosystem and the living rights of citizenship. It is not only natural but also human diversity that must be extensively conserved or recovered, because the content of the rich nature and society worthwhile being sustained is diversity. In such a diverse system, any member is recognized as the participant to be relied on rather than as the object to be exploited or conquered. The association-style 'kyosei' strategy must respect other parts and its link with them, and also utilize fully traditional wisdom and moral to avoid overuse of resources artificial as well as biological. Unless we think seriously that the unchecked growth of a species in nature or a nation in humankind will eventually lead to 'kyoshi', or common ruin, we will not be able to survive the implications of the eco-systeem or socio-system crisis.

System resulting in ‘kyoshi’

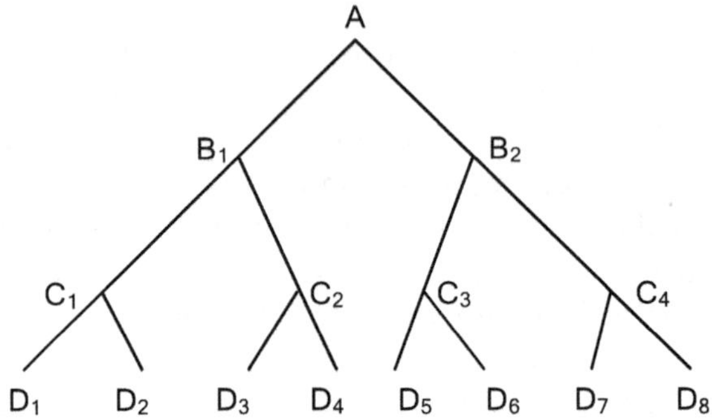

Fig. 3.1. Condition of Death: Non recurrence-non cycle, in which A→B→C→D.

System sustaining ‘kyosei’

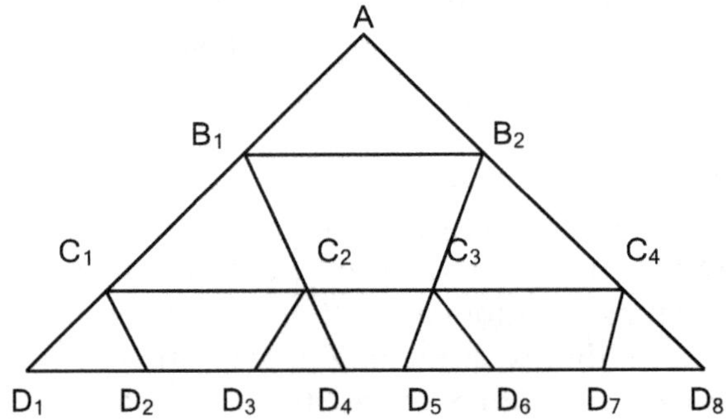

Fig. 3.2. Condition of life: Recurrence-Cycle, in which A→B→C→D→A.

6 Conclusion

Our consideration of ‘kyosei’ sets up a different model from the mutualism-type ‘kyosei’ one, based on eco-semiotic interpretation of the biological association. Our model shows that the semiotic correlation characteristic of the social and natural ecosystems determines how the former should participate in the latter and manage or care for its systemic behaviors; and, however seemingly free and independent at its own level, the socio-system globally repeats the patterns and cycles of 'implicate order' or 'One'.

So the major premise should be: first, our sustainable socio-system must depend on conservation of the natural ecosystem; second, we should convert from the idea of increasing economical growth as the only highest value to that of raising our quality of life under the ethical action plan of sustainable development sustainable; last, our urgent task is to set a good condition for everyone to take part in semiosis creatively and to share the mechanism of how to generate, transmit, and develop creative knowledge, while also being cautious that such a creative knowledge may be able to generate the undesirable results initially unexpected. Quite unintentionally, changes in artificial ways can often affect the natural environment, so we must take into consideration the concomitant results deriving from creative semiosis and prepare for the measure that will allow to quickly deal with such negative results.

Masanori Funakura, Division of Mathematics and Human Science, Graduate School of Tokyo University of Agriculture and Technology

Masayuki Horio

'Symbiotic technology' in the early 21c. ordeal

Abstract:

Now, after some three hundred years, we are in a period of serious review and reconstruction of our predecessors' achievements, the modern technology and the present civilization based on it. To obtain a perspective for the survival and sustainable development for the third millennium it should be of some value to examine the basic nature of modern technology by trying to find out the essential methodology to reconstruct it from the present egocentric state to the symbiotic and sustainable.

In the present article, first the definition of technology is discussed converging to a new and comprehensive understanding of the nature of technology. From this, conditions for making socio-technical systems more consistent which are necessary for sustainability and symbiosis are derived. In relation to the above understanding the effective social and academic actions to cope with the present situations are discussed. Some recent activities of the author's group ESTeC in socio-technological development are also briefly presented.

Keywords:
essence of technology, sustainability, utility technology, reforming, collaboration

1 Introduction

The human civilization is now in the midst of industrialization. The industrial revolutions and the serious changes in social structure induced by them swept major western countries and Japan from the late 18th century to early 20th century. If we look at the world history not from the vantage point of the west, we may feel reasonable to discriminate industrial revolution from technological revolution. In the western history industrial revolutions were almost always along side with technological revolutions. In these countries, social revolutions and/or reforms of governance to a nation state were conducted in their early period to make them ready to any significant technological/industrial revolutions. In the terrain of developed or well ready countries a significant technological revolution may immediately bring about an industrial revolution. However, it is not always so depending on the industrial readiness, which is relevant to the social and economic systems. Once reforms in governance necessary to provide such readiness are fulfilled, modern industrial revolution can be triggered to penetrate also into developing countries. China, the country of the largest population in the world, is now in this process. This implies that, as I stated in the beginning, the world is now in the middle of the historical 'industrialization' in terms of population involved within it. These 'latecomers' to the industrial revolution have of course the benefit of utilizing the top sum of achievements from the technological revolutions. Starting from the first revolution of motive power, the subsequent revolutions continuously introduced new socio-technical components such as electric power, fuel powered transportation, refrigeration and heat pumping, plastic materials supply, artificial means of communication and information technology, etc. into society. During this process the social system for knowledge development and transfer has been built up and technical innovations once lead by craftsmen were taken over by well educated engineers and scientists. National budgets have become the major source of scientific/technical education and technology development. The New Industrial Estates are of course keen to introduce a strong program of professional education and technology developments.

Although it was only a coincidence, right before China's participation in the world's modern industrial revolution the fate of the continuous technological development was perceived by leading thinkers. As well known, it was then followed by the oil crisis and the scientists' admonition of global warming. Essential problems lie in the present lifestyle in the massive production of wastes and chemical pollutants which have become in all respects clear to people. Even the vague idea of sustainability also has been rapidly accepted by the public.

Nevertheless, it is a clear issue that the present socio-technical system or civilization that stands heavily on fossil resources cannot survive another millennium. One can expect technological developments of nuclear fusion and plutonium cycle to provide a sounder basis for the technological society. Here, it is also true that any technology dreams may fail. If the probability of the failure is not negligible, and I believe so, we have to prepare ourselves for situations of no more cheap fossil fuel energies. Even if nuclear fusion or the plutonium cycle becomes technologically realizable, the society that hold such a technology could be a highly security controlled one, which may not be comfortable at all for humans as expected by an anti-nuclear activist Jinzaburo Takagi (1994). Setting aside the issue of accepting the present nuclear policy or not, such a long term design of comfortableness of life should be carefully investigated. Anyway, the possibilities of energy and environmental disaster should give a huge threat to the present civilization. But probably because of the ramified structure of scientific research and education, even professionals are yet to seriously recognize the inexorable developments of future energy crises (*pl.* since there should be many critical stages until we have spent out a major part of our fossil fuel energies). We will face having to make choices in our fundamental demands and desires and our lifestyles will have to be moderated in the future due to the limitations in energy resources and environmental capacities although the main framework of the modern technological society will be preserved.

This leads to the necessity of reforming the present civilization into a sustainable one. But this will not be done 'without attacking the premise of the dominant social paradigm' (from the definition of environmental reformism by Bill Devall, 1980). Since the necessary reform should be done from the depth of technological design principles, I, as an engineering scientist, contend that attacking it only from the tenets and inspirations cannot change the technology from the inside. Moreover, people are the everyday users of the present technology. To obtain the majority's support it is significant to search for the solution together by talking to the public both inspirationally and technologically. As the term ecology acquired its present benign reputation in the society through developing quantitative evaluation power such as energy economy and life cycle assessment, no matter how primitive they are at present, the attack should also be interpreted into a more quantitative technical language.

In Japan, the modernization and industrialization proceeded to the extent that the generation rate of waste cans from beverage is as high as in the Unites States. The significant Americanization was made during the late 60s to the mid 70s. We are already two full generations separated from Japanese beliefs of harmony with our beautiful nature and traditional culture that the early modern Japanese counted on to balance their spirit during the modernization all the way from Meiji Restoration to WWII. The 'attack' should be well organized to be both of deep spiritual impression and of technological rationality.

The objective of the present article is to try to find a link between engineers' practice and the public as well as nature with which the technology can grow out of the present anthropocentric state to the symbiotic.

2 What is technology

2.1 'Definition of technology' dispute in Japan

In early 1930s Japan was about to enter a new era of military operation in northern China after the bubbling economy of post WWI and the recovery from the turmoil of the Great Earthquake of 1923. It was that time (1932) when Japanese leftists organized Materialism Study Society and adopted the technology study as one of the main topics of their polemics. Aircraft technology was also about to take off with the Navy's Aviation Technology Independence Program (1931). To construct science and technology infrastructure independent from western powers was indispensable to anybody. Reacting to such flavor of the period and triggered by overseas arguments on technology in Soviet Russia and their 1st five year plan (1928-33) as well as by those in Germany (e.g. Oswald Spengler), their study lasted until the dissolution of the association (1938) due to government's oppression and the mass arrest of leftists in Mantetsu (South Manchurian Railway Company). The controversy was however, inherited by the post WWII leftists in a dramatic way that a quantum physicist Mitsuo Taketani's article on Technology was published in 1946 as a copy from Special High Police records at his interrogation. In this article he attacked the old Marxists definition of technology, *'technology is a system of labor instruments'*, and submitted his essentially disparate counterproposal, *'technology is an intentional application of objective natural rules in humane productive practices'*, which was welcomed widely by scientists and engineers of that time because it matched their everyday practices. The post war dispute became silent in early 70s after Japan went through the university crisis with student riots and serious anti-pollution movements, which came to an end with the enforcement of 14 anti-pollution acts of 1970 followed by the Limit to Growth of the Club of Rome (1972) and the oil crisis of 1973 (a good summary of the dispute was done by Seiji Nakamura, 1975).

As a Ph. D. student at that time, I spent some of my free time to revisit old literature and derived my own criticisms, which can be summarized as:

1 Categories such as labor and means of production relevant to the microscopic labor processes with economic discrimination between those with ownership for the means of production and those who don`t are not appropiate for defining the essence of technology.
2 Technology is inherent in an existing production system but not congruent with it. If there is no distinction, we do not need the term technology.
3 From these two old Marxian definitions, one faces immeditae difficulties to deal with chemical and biological issues that can be either object or instrument of labor time and such change is all trivial for a technology consisting of them. On the other hand it faces another difficulty with the issue of labor quality, which is in many cases essential for realizing a new technology. No wonder that in Soviet Russia 'facility fetishism' overwhelmed and efficiency was thrown out.
4 Taketani's 'application' is the microscopic process of technological practice, which is not at all sufficient to define the outcome of an integrated technological system.
5 Taketani's definition cannot really include the potential failure and potential unintended success that are essential in the existing technology.

As illustrated in Fig. 1, technology is contained in an existing system which is an integrated sum of previous technical practices. Since it is a result of previous human work, it can be either advantageous or harmful to people living in the present time. Even if it is a comfortable system to the present people, it can be an apparatus to make them obsolete through self-domestication.

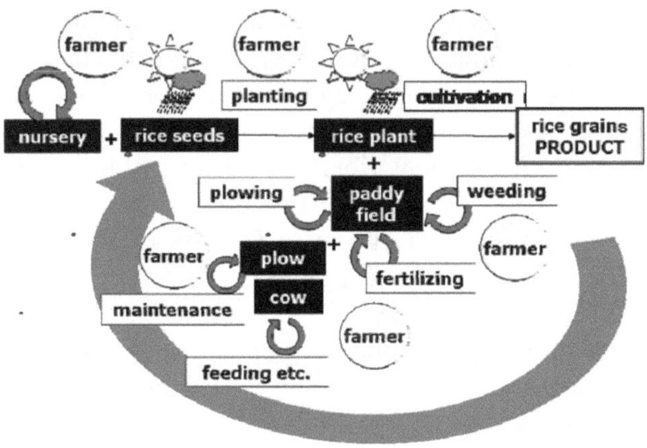

(a) Individual labor processes in the total rice production process

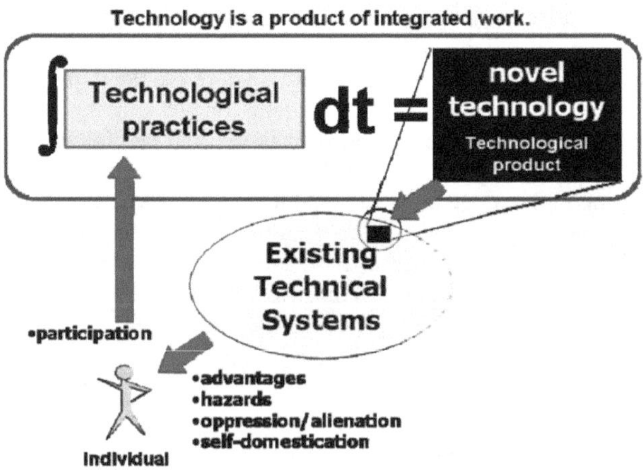

(b) Existing technical systems developed in the past and an individual living now

Fig. 1. Illustration of the integrated natures of technology

2.2 Definition of technology in its narrow sense

The definition I submitted at that time (Minami, 1973) was *'technology is the mechanism in a productive force'*.

With this definition we can look at existing technologies in their reality, in their integrated form, where all the previous applications, failure and success experiences, even those still unrecognized lucky tricks are integrated. There we also see the integration of all the necessary parts, feed stocks, labor and labor-, safety- and emission-imperatives. With the presence of unrecognized aspects in technology there arises a good reason to investigate newly introduced technologies in detail for further improvements, new inventions and theoretical generalizations as are daily conducted in engineering schools. Since it is a mechanism, it can be expressed by technical and scientific terms and can be transferred by drawings and can only be protected by the social patent system and knowhows.

Other intriguing traits of technology can be found by taking the case of cogwheels for example. The technical principle of gears was already known in the period of Archimedes. The principle itself can last thousands of years and has been improved even before the industrial revolution as the gear cutter invention of a Swedish engineer Christopher Polhem (1691-1751). However, it was only after the establishment of scientific theory of gears that the rational design principle was generalized by taking into account the bending strength of gear teeth. The gear tooth curve is nothing but a geometrical curve but the materially realized curve is different in the sense that it can never correctly follow the demand of geometry in a microscopic level, that it may go through plastic deformation or erosion during operation and/or that it may be sintered with others by rusting. Thus at a period during aging there arises a phenomenon where we can see gears but they cannot show their expected performance or they become even completely motionless. Then, gears exist physically but practically they cannot perform. As we can read information from a dead body, we may be able to obtain necessary data to specify the gears but that they cannot move remains a fact. Thus, the relationship of technologies and a production system, including machines, labor and raw materials, is to some extent similar to life and body issues.

2.3 Further extension of the definition

The definition of technology discussed above is only the one in its narrow sense. We can extend the concept to a much wider perspective by viewing its social aspects and its behavior in the time domain. Laws, regulations, technical standards and all the social installations helping a technology being suspended in the society should be necessary companions of a technology. If any materialized existence has its life time, its reproduction or re-procurement is a key issue for the being of a technology. Reproduction of technical parts and maintenance are the basic requirements. Moreover, we may introduce all infrastructures for technology re-procurement such as education for labor reproduction including engineers', reproduction of the economic power for purchasing necessary goods to keep the process going and political and legal systems for the milieu of the relevant production.

In the philosophical disputes over machine and organism issues, it is often stated that a machine cannot reproduce itself but an organism can. However, even an organism is not reproducing itself automatically. If we look at the repairing system of an injured cell, we can find all the complicated cooperation of the lymph system and other organs. This means that reproducibility does not immediately mean autopoiesis. A system is Autopoietic (Maturana, 1980) if the bits and pieces of which it is composed interact with each other in such a way as to continually produce and maintain that set of bits and pieces and the relationships between them (by Tom Quick). We should not forget that we have our leukocytes and/or lymphocytes in our own society, i.e. maintenance engineers, who repair our machine organs. Machines and production systems are parts of our organism named 'society'. We need a macroscopic view point to define technology.

Another extension of the concept of technology is that it does not remain at the machine and/or factory scale. The social mass and energy handling systems as well as waste management systems, i.e. the utility systems, also contain mechanisms with which they can be effective. Thus we can introduce the concept of 'utility technologies', which are necessary to make out our lives in the society. Now in the environmental era, the utility technologies should include all the safety issues and controls over individual industries' activities. Good example can be found in an old industrial city named Kita (north)-Kyushu City, where Japan's 1st blast furnace was erected in 1901. The city is now busy reforming itself into an Eco-town, with a recycling production center dealing with a variety of industries including iron/steel plants, cement plants, electric power plants and many others by establishing Environment and Industry Division responsible for both industrial promotion and environmental control.

2.4 Technology and self organization

As already discussed a little in the above arguments, technology can be viewed in the light of the life and body analogy. It is interesting that a gear system can be meaningless if they are abandoned and left in nature. Technical systems are the product of self organizational effort by human beings. Without it, systems can be meaningless or in other words, it looses a function that can take place with the human being who understands how to use it. Of course, as crows in Japan love to steal iron wire hangers to make their nests, some technical products can be appreciated by other beings. However, crows remain as crows even though they use human products. Equally, pig organs can be implanted in human bodies without changing the human nature.

Technology development is the progress in the degree of self organization. The decisive factor in the technology development is the presence of a center as in the case of higher forms of life. With a strong will of the center the rationality and efficiency of a technology would be promoted. If the will of the center is not sufficiently strong or in confusion, a high degree of its completion would not be achieved.

Viewing the process of a technology development from this window, we can evoke the importance of autonomy. How its motivation was formulated, how development efforts were organized and examined and so on are always critical in any collaborative work. Taking the automobile technology as for an example, we know Toyota's success and Nissan's loss and recovery. Among companies competing for new models and new technologies, those that can organize themselves into a group of strong self control and challenging spirit for development tend to produce well thought out and consistent designs and novel inventions.

Good commodities thus overwhelm the market as they are adopted by the majority of people and become socially accepted products. We remember quite a few success stories from Edison's lamps, Ford's Model Ts to the recent Sony Walkmans, Nokia mobile phones and Japanese digital cameras, which brought almost irreversible changes into contemporary lifestyle. These became socially incorporated technologies. However, the essential nature of their creation is based on private ownership.

2.5 Insufficient self organization in public/utility domain

Assisting commodity technologies, utility technologies have been developed to supply water, gas and electricity and to provide services of mailing, telecommunication, transportation including railroad, bus, taxi and road connections, security and waste disposal. Some of the systems are owned privately and some by the public. Although all of them are to some extent under control of municipal and national governments, the degree of self control is becoming more doubtful particularly recently when the paradigm of 20th century's 'everything convenient and labor saving is good' has become questionable. Selling more gas and electricity can no more be a leading principle. Developing wider road and highway systems may make the traffic jams even worse, causing more jams, pollutions and fuel depletion. Just constructing more incineration stations to cope with rising garbage production simply fosters more generation. Organizing people's discussion over the solution to avoid waste, reduce garbage and promote unobtrusive lifestyles is even more effective. If more energy can be recovered from garbage by combusting them at power stations rather than incineration stations presumably with more strict separation of hazardous wastes from it, people's participation becomes a key issue particularly in terms of their responsibility for strict separation at their home. At the same moment the industrial partnership becomes necessary to realize such social systems. These are directions quite different from those derived by the modern principle of convenience and labor saving but they are still in quite a primitive stage.

Government controls both on the private sectors and on publicly owned sectors are in dubious conditions affected by stiff sectionalism as well as by antiquated laws and by-laws that can be much apart from technically rational imperatives. To ameliorate such situations democratic intervention is one of the necessary factors, as discussed by Andrew Feenberg (1999), but it is not at all sufficient. Since people are the beneficiary of utility technologies and are paying for the services, but ultimately for the preservation of them, they have interests as well as rights to propose any changes. However, if the technological solution to a given situation is not well sought out, the intervention cannot be productive no matter how democratic it is. On the other hand, the technical experts in the private sectors have their own interests that may not be the same as the relevant people's. Things are the same for academic researchers since they are more or less controlled by research funding based on the prevailing paradigms that may not be effective for solving local residents' issues. Here, I just would like to stress that in the case of utility technologies the center for its rational development, or for its appropriate self organization, is still incomplete because of the situations described above (see Fig. 2).

3 Toward sustainable and symbiotic technology

Nevertheless, conditions for making the present socio-technical systems more consistent are now improving rapidly. As illustrated in Fig. 3 we have been constructing a common agreement on sustainability, which should be one of the leading concepts to unite people both in local and in much wider terrains. To achieve substantial progress in the remodeling of utility systems we first need collaboration among regional people, their municipality, experts and private manufacturers and service providers. Interestingly, in a matured society we have quite a few retired experts in many fields who can bridge lay people in the region with external experts. Second, to orient the collaborative efforts toward a creative and constructive direction we need mechanisms to engender trust and consensus among collaborators. For such requests the recent information technology can provide platforms for planning based on LCA (life cycle assessment) and MFA (mass flow analysis), which can be utilized in the collaboration making the consensus process much intensified and trustable. The third component we need is the spiritual integration for making a strong headquarter in the regional society, where the historical prides, cultural uniqueness as well as the potential roles of the regional society in the national and global terrains should be combined.

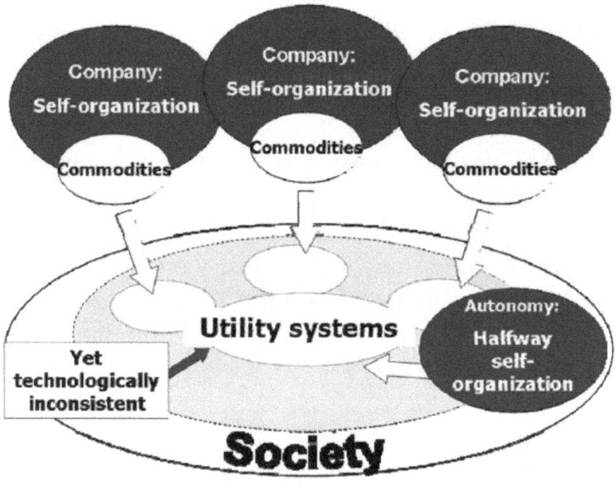

Fig. 2. Illustration of the cause of technically inconsistent utility systems

If local nature conservation/creation issues as well as reforming the regional social mass/energy metabolism issues are incorporated into the tasks of the collaboration, a sure ascending step would be designed for our sustainable future. There I believe even deep ecologists' ideals can be carried on in a more quantitative and step-by-step manner.

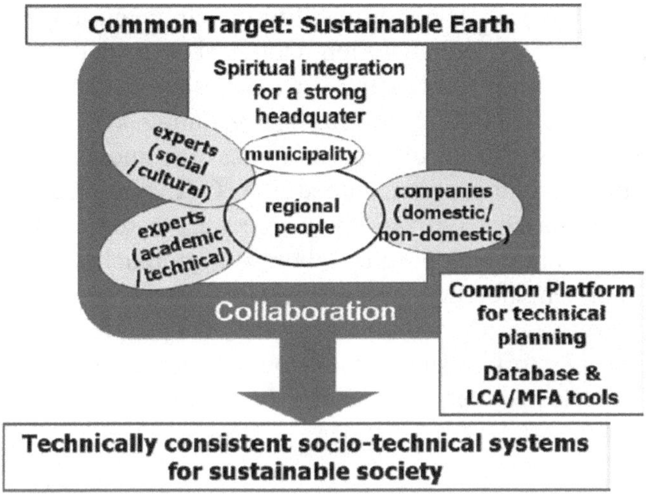

Fig. 3. Necessary components for consistent socio-technical system construction

4 Concluding remarks

To the question if technical systems destructing our environment as well as bio-diversities should be destructed, I would answer no. I would rather disintegrate them and reconstruct them into better ones. Actually, most of the parts of technical systems can be utilized for both good and bad. They are our treasures anyway. We human beings are created through anthropoids' technical practices and can never live out of technological systems. It is the fate of technology to always seek common conveniences, whatever they are, and also compatibility. However, the most embryonic field of the modern technology is the group of social or utility technologies whose reforming is now our serious task for sustainability.

In my university, we have been redefining the role of technical universities. Previously it was simply to lead our nation for industrialization. Lately, it has been said that we have to lead technical innovations for maintaining economic competitiveness. Collaborations between academia and industry have been promoted by the national government. However, for sustainability development and renewable energy utilization, the collaboration should be extended to local municipalities and relevant associations. Universities also have to be keen in developing public platform for trust and planning, keen in examining technical concepts in terms of appropriateness in regional societies and so on. In our five-year program named ESTeC (Evolution and Survival of Technology-based Civilization), where I have served as a leader since 2002, we are developing collaborative approaches among technical (both in engineering and agriculture), social and cultural researchers as well as with municipalities, regional people and industries, which should be our first step to make ourselves ready for the early 21 century's ordeals in water, food, energy and culture crises.

Bibliography

B. Devall, 'The Deep Ecology Movement', in *National Resources Journal*, 20/2, 1980, pp. 219-313.

A. Feenberg, *Questioning Technology*, Routledge, London and New York 1999.

H. Minami, 'Tasks of Engineering Sciences and the Science of Technology', in *Journal of Japanese Scientists*, 8/2, 1973, pp. 6-11.

S. Nakamura, *History of Disputes on Science of Technology*, vols. 1 and 2, Aoki-Shoten, Tokyo 1975.

T. Quick, <http://www.cs.ucl.ac.uk/staff/t.quick/>.

O. Spengler, *Der Mensch und die Technik. Beitrag zu einer Philosophie des Lebens*, Munchen 1931.

J. Takagi, *Future of Technology – A message from 2041*, Iwanami Shinsho 365, Iwanami-Shoten 1994.

M. Taketani, 'Theory of Technology', in *Problems in Dialektik*, Rironsha, Tokyo 1964, pp. 171-191.

Masayuki Horio, Department of Chemical Engineering, Tokyo University of Agriculture and Technology

Shuji Ozeki

'Kyosei' and sustainable society: from the viewpoint of environmental philosophy

Abstract:

The Japanese word 'kyosei' (or symbiosis) has been discussed by famous intellectuals in various fields and by the public in contemporary Japan since the end of the 20th century. It is generally well known that, 'symbiosis' is a word used originally in biology. However, in contemporary Japan it is insisted upon as a philosophical idea concerning the human-nature relationship and the human-human relationship, just as the words 'struggle for existence' used by Darwin in his evolutionary theory from the second half of the 19th century to the first half of the 20th century (Social Darwinism).

The word, 'kyosei' literally means that the different live together, there are important differences in meaning according to the various groups disputing theories of symbiosis. I will attempt to examine the differences and classify them into three types, and through the process my own understanding of 'kyosei' will be clarified. Based upon my understanding of the word, I will discuss its meaning especially with regards to 'the symbiotic relationship between humans and nature', considering also the changing of the modern view of the human-nature relationship to so-called 'Environmental problems'. This also means that we should think of the complex mutual action between the natural and the artificial and its effect on the human-nature relationship. Finally, the importance of 'communicative attitudes to nature' in connection to 'kyosei' is discussed for the formation of Sustainable Society, showing that philosophical idea of 'kyosei' also transcends the confrontation of 'anthropocentrism' versus *'physiocentrism'.*

Keywords:
kyosei, sustainable society, environment philosophy, naturalness of human being, communicative attitudes to nature

1 Introduction

In contemporary Japan, the word 'kyosei' (or symbiosis) has been discussed frequently in various fields and by the public during the past 20 years or so. It is generally well known that, 'symbiosis' is a word used originally in biology. However, in contemporary Japan it is insisted upon as a philosophical idea concerning the human-nature relationship and the human-human relationship. Particularly, a 'kyosei' boom occurred in the journalism and mass media in the late 1980's and has continued until now. The discussion started with 'symbiosis between humans and nature' and developed to a variety of symbiotic relationships such as 'symbiosis of races', 'symbiosis of cultures', 'symbiosis of genders (sexes)', and 'symbiosis with the disabled'. However, the word 'symbiosis' has often degenerated into something that attaches an aesthetic tint to commodities similarly to the word 'eco-friendly', inviting no small repulsion. In these circumstances, some critics maintain that we should avoid the term 'symbiosis'.

However, the term 'symbiosis' is considered to be one of the key words that symbolize the turn of the epoch in the modern times. In addition, this term is particularly interesting in that it resembles 'the struggle for existence', a word once used across fields of natural and social science as a key word of the time. From a macroscopic view, the concept of 'the struggle for existence', which was the keynote in not only biology but also social sciences and man's relationship with nature from the mid 19th century to the mid 20th century, appears to be giving way to the concept of 'symbiosis'.

However, the present time may be described as a time of transition, in which the concept of 'symbiosis' and the concept of 'the struggle for existence' conflict with each other, considering the fact that the renaissance of competitionism is strongly advocated under the banner of 'neo-liberalism' and with the slogan of 'global competition' after the collapse of the former Soviet Union and the East Europe, and with the difficulties confronting welfare states. This situation also seems to nurture conflicting arguments and understandings concerning the concept of 'symbiosis'.

I am an enthusiastic advocate of the concept of symbiosis as a philosophical idea for the realization of a 'sustainable society', and, from this standpoint, I would like to remind the readers of the following points concerning the significance of the concept of symbiosis.

First, it must be understood that a 'sustainable society' is a society with an aim to overcome environmental destruction and wasting of resources as a basis, but it should not be a controlled society based on narrow environmentalism. It must not be a society that runs counter to environmental justice or suppresses cultural diversity, let alone an eco-totalitarian or eco-fascist society. Moreover, a 'sustainable society' should not interfere with unique self-realization of each of its members.

Thus, if the concept of 'symbiosis' is a concept that leads to the realization of a sustainable society, it must be asserted not only in the human-nature relationship but also in human-human relationships.

In contemporary Japan, the participation of so many people in the discussion of 'symbiosis' seems to be causing utmost confusion. Therefore, I would like to present my understanding of the reality of 'symbiosis' by classifying the discussion into several types. All discussers of 'symbiosis' direct their interest to the relationships among differences and different entities, but their contentions may even be completely opposite.

2 How to understand the reality of the concept of symbiosis

2.1 Three patterns of the concept of symbiosis in contemporary Japan

As mentioned earlier, 'symbiosis' has often been a popular topic in contemporary Japan, but it has been discussed, characteristically, from different, and sometimes seemingly conflicting, viewpoints. I would like to classify these viewpoints into 3 types and term them as follows: 'symbiosis of sanctuaries', 'competitive symbiosis', and 'communal symbiosis'. To mention this in advance, when they are related to the viewpoint of the *modern times*, symbiosis of sanctuaries is considered to have a pre-modern orientation, competitive symbiosis to have a modern orientation, and communal symbiosis to have a post-modern orientation. In the following sections, I would like to discuss each of these types by referring to its representative advocates.

First, 'symbiosis of sanctuaries' is represented by the views of Kisho Kurokawa, a famous architect/philosopher. In a considerably early stage of the dicussion, he discussed the importance of symbiosis in a systematic way by writing a book titled *The Philosophy of Symbiosis: Lifestyle for survival in the future* [Tokumashoten, 1987]. According to Kurokawa, 'the concept of symbiosis is an attitude to respect each other's sanctuaries' [*ibid.*, p. 97]. Since Kurokawa's contention was made in the circumstances of an upsurge of economic frictions between Japan and the United States, it was related to his opposition to the pressure by the United States for greater freedom of trade, and he brought up his 'sanctuary theory' in this connection, stating, 'I consider the Emperor system, rice growing, *sumo* wrestling, *kabuki*, and the tea ceremony (*sukiya* architecture) to be Japan's sanctuaries' [*ibid.*, p. 98]. The following are his remarks about the Emperor system.

> I think that the present Emperor system, or 'the Emperor as a symbol', is a part of Japanese culture that embodies Japanese history itself and the pride of the Japanese. It is nothing other than a sanctuary, and I cannot approve the intentional emphasis by the Emperor that he is an ordinary person. *Kabuki* and the tea ceremony, which are patrimonial disciplines, the *sukiya*-style architecture, which is possible only by skills transmitted over generations, and *sumo*, which is not a sport but is a ritual associated from its very origin with the Emperor system and rice growing, should also be considered typical Japanese sanctuaries.
>
> A sanctuary is regarded as a sanctuary not on the basis of scientific analysis or international rules but because of its aspects such as the unintelligibility, mysteriousness, a source of identity, and a source of cultural pride [*ibid.*, p. 100].

Such concept of symbiosis, which rejects communication for mutual understanding from the beginning by bringing up a 'sanctuary', can be described as a pre-modern and conservative type. Nevertheless, his inclusion of 'a source of identity' and 'a source of cultural pride' in the concept of symbiosis is worth noting. Takeshi Umehara, an influential philosopher with good sense, who emphasized 'philosophy of symbiosis and circulation' on the basis of Japanese traditional thoughts including Buddhism, was among the advocates of such a conservative concept of symbiosis. [Umehara, 1996].

However, conservatives who became eloquent in defending neonationalism in the face of the revelation of the problem of 'military prostitutes' and other war-time atrocities since the mid 1990s appear to have no room to discuss 'symbiosis' any longer because of their belligerence against foreign countries. Instead, emphasis on the development of national competitiveness to survive the era of globalization appears to be fostering a frivolous reversion to 'the struggle for existence'.

About the second 'competitive symbiosis', Tatsuo Inoue, a liberalist legal philosopher, presents a typical argument. Inoue ardently advocates the concept of 'kyosei' as a new concept that regulates new human relationships in the Japanese society. This proposal of Inoue is based on his critical attitude to problems concerning the modern Japanese society. With this attitude, he considers liberalism to be a more important viewpoint for criticism than democracy at present to challenge the tendency toward homogeneity and exclusion of heretics, which are regarded as characteristics of the Japanese society, because the basic concept of liberalism is 'symbiosis of heterogeneous and diverse independent personalities'.

> *Kyosei* in the sense that we use the term is a pattern of social bonds open to things different in nature. It is not friendly coexistence and co-prosperity within a closed society but it is social bonds through which people with different lifestyles mutually approve free activities and opportunities of participation and actively build up mutual relationships [*Adventure of Kyosei*, The Mainichi, 1992, p. 24].

Therefore, Inoue is critical to the system of symbolic Emperor, because he considers that serious problems with the symbolic Emperor are that it promotes the tendency toward homogeneity and makes *kyosei* impossible.

> Why does the Emperor system serve as pressure that produces various impingements of human rights? *Invisible minorities* presents a key to an answer

> to this question. The Emperor system alienates the Emperor as a specific individual but attaches irresistible centripetal force to the Emperor as a public and established existence. In the core of this centripetal force is a force that makes minorities invisible. To oppress heterogeneous minorities to make them invisible and to structuralize such oppression, this force makes the very fact of oppression invisible. (omission) The force that makes minorities invisible is a force that eliminates practical and/or latent heterogeneity and a force that withers the personal independence of not only particular minorities but also all individuals. This force that makes heterogeneity invisible and erodes the freedom of being different is the essence of what is called the 'myth of a homogeneous society' [*ibid.*, pp. 116-117].

Thus, the concept of kyosei of Inoue and others challenges the background of traditional discriminatory mentality that persists and is reproduced in the Japanese society. Therefore, in comparison with Kurokawa's concept of symbiosis, this concept may have a contemporary significance in modern Japan. To me, however, Inoue's critical awareness of the present excessively 'competitive society' seems weak compared with his criticism against the homogeneous society, because his concept of symbiosis is based on positive affirmation of the market principle of competition. Certainly, Inoue maintains that there are two types of competition, and while he reproaches competition to 'catch up and get ahead' toward a given goal, he praises true competition for diversified goals as a source of 'richness of humanity'. However, he fails to see the reality that excessive competition of any form may become 'the struggle for existence' in contemporary Japan, where social security is defective and relationships of mutual aid become sparse.

Therefore, his concept of 'symbiosis' is good as long as it aims to overcome the 'tendency toward homogeneity and exclusion of heretics' observed in the symbolic Emperor system, but it also contains a competitive principle that rejects competition that promotes homogenization of the society but encourages competition that advances heterogenization and individualization. In short, Inoue's concept of 'kyosei' can be called a concept of 'competitive symbiosis' that has a negative attitude to communalism in general as well as to conservative 'communalism'.

In this respect, a few philosophical authors, such as Toshiharu Takeuchi, Satoru Saishu, and Kohei Hanasaki, who are interested in *the socially disadvantaged* including the disabled and ethnic or cultural minorities, expressed some sympathy toward Inoue's argument about symbiosis but, at the same time, understandably voiced discomfort and doubt in unison. In a competitive society with the law of the jungle, talks about 'symbiosis' by the strong often lead to concealment of oppression behind the façade of equality. In such a situation, pursuit of practical equality between the weak and the strong from the viewpoint of justice of distribution is indispensable for the realization of true *symbiosis*. Therefore, a concept of 'symbiosis' that contains resistance to competitionism by union and social solidarity of the weak is considered to be important.

I would like to tentatively call such a concept of symbiosis based on *communalism* 'communal symbiosis'.

2.2 Concept of 'kyosei' based on communalism

In the previous section, I proposed my concept based on *communalism* as an alternative of the concept of symbiosis based on *competition* by reviewing various types of concept of 'symbiosis'. I also suggested that this concept of symbiosis can be related to problems of overcoming both restorationism to the pre-modern times and market-centered modernism. Symbiosis should also be discussed in connection with the management of the capitalistic modern times with its laws of the jungle.

Such a concept of symbiosis based on communalism may be defined as an extension of a critical antithesis to the concept of community in the mainstream of the conventional socialistic thoughts in consideration of the fact that it in part promoted the tendency to homogeneity often in universalistic manners similarly to modernistic thoughts in general as they were trapped by paradigms of the Enlightenment while it was originally developed to overcome the capitalistic modern times.

This is because this third type concept of symbiosis based on communalism is not simply juxtaposed with the first type or second type concept as their simple antithesis but admits their positive aspects and preserves them as its own moment. Thus, following the first type concept, it does not consider it possible to pursue completely transparent and intelligible beings as the modern Enlightenment did, but it attaches importance to the approval of and respect for beings that is unintelligible at present while it continues to pursue the intelligibility. Also, following the second type concept, it recognizes the aspects of struggles and confrontations in symbiosis as moment; symbiosis is not a friendly company. The third concept maintains that, despite these struggles and confrontations, there is communalism deeply rooted in humanity at its base.

Human beings are innately *communal beings* as evidenced by the presence of collaboration and languages without having to refer to the Aristotelian term of *zoion politikon*. The communal nature of humans and acts of their mutual assistance based on it are considered to be deeply rooted in the essence of humanity.

However, as such innate communal nature of man has become less obvious under historical conditions, atomized and isolated individuals have begun to engage in selfish competition, and the struggle for existence prevails as if it were a permanent character of humans in the system of laissez-faire capitalistic market economy. In a society where exchanges of commodities prevail and money has a dominant power, the relationships between isolated human beings appear as relationships between objects. The modern times observed a phase of the human history of radical destruction of natural communities through the universal spread of commodity exchanges. As observed by the economic anthropologist Karl Polanyi from Hungary, 'commercialization of nature and man', the dominant trend of the modern period, is a very unusual phenomenon in the 4-million-year history of the mankind.

As briefly observed above, human beings have innate communalism as part of *human nature*, and their communal values derived from this nature are considered to have played important roles in most of the 4-million-year history of the mankind. As mentioned above, capitalistic modernization, which began only a few hundred years ago, obscured these communal values by atomization of members of the society and their struggle for existence while it induced strong awareness of personal values of individuals on the citizen level with the advent of modern individualism and liberalism. For this purpose, the realization of not only formal but also practical equality is important, and it is made possible by restoration of communal values. Such human history factors must not be overlooked in the background of the contemporary dispute over the concept of symbiosis.

In a sense, *symbiosis* as I understand is considered to be a principle which may allow us to overcome with communal values the abstractness of individualism and egalitarianism brought about by the modern society. This is to attempt to qualitatively convert individualism by a manner that approves irreplaceable values of the individual identity in the modern times on the basis of the recognition that communal values are indispensable for human beings. It also attempts to substantiate egalitarianism that maintains that all humans have equal values in the diversity and heterogeneity of human beings rather than to confine it in the modern abstract formality.

In this perspective, the concept of symbiosis based on the *communal nature*, i.e. the concept of 'communal symbiosis', is considered to be a position that attempts to positively approve personal values, which are trophies of the modern times, and to realize practical equality by surmounting the modern formality by building on communal values based in the natural origin of human beings.

'Symbiosis' may be discussed with regard to the following 3 major aspects.

1 Symbiosis of life
2 Symbiosis between humans and nature (relationships between individual human beings and animals, relationship between the human world and nature, etc.)
3 Symbiosis of human beings (individuals, groups, races, etc.)

The above discussion of the concept of symbiosis primarily concerns *3* 'Symbiosis of human beings'. *1* is evaluated primarily in biology. Therefore, in the next step, let us reflect on *2* 'symbiosis between humans and nature' on the basis of the discussion about *1* and *3*. It will, by its very nature, lead to deeper contemplation of 'environmental problems'. Evaluation of symbiosis between humans and nature relates to more complex evaluation in the framework of relationships between humans and its environment.

On the one hand, nature as an 'environment' must be understood also as already socialized and artificial nature. On the other hand, the radical meaning of 'naturalness' in the social and cultural human being must be contemplated. 'Symbiosis between humans and nature' must be evaluated in such complex interrelations, and it is closely related also to 'symbiosis of human beings' through problems of social systems. I would like to discuss crises of the modern society and their conversion to a sustainable society by evaluating these topics.

3 'Environmental problems' and 'human naturalness'

3.1 Philosophical meanings of 'environmental problems'

Many people presently feel that the future of human beings is veiled in dark clouds rather than is rosy. Not a few people would become profoundly skeptical when they think whether the human race can dissipate these dark clouds in the 21st century. Among these dark clouds are so called 'environmental problems'. The destruction of the global environment is increasing in its seriousness to such a degree that many scientists appear to feel that the situation is already irreversible. Such a notion is intensified by the realization of the graveness of the effects of environmental hormones in addition to air pollution.

While the threat to the life including man is talked about amid the aggravating environmental issues, the technology to manipulate life has been advanced remarkably to the point where organ transplantation and artificial insemination have become commonplace, clone sheep have been produced, and even cloning of humans has become a topic. Manipulation of life might appear to shed light to solution of various problems that have been out of reach of human beings. Today, however, more people seem to feel uncanniness of manipulation of life than express hopes for its further development. They have the concern that, in the long run, manipulation of life is a kind of destruction of life by the technology. From such a viewpoint, the problems of manipulation of life become inseparable from 'environmental problems'.

While these problems related to the environment and life are inevitably associated with 'darkness', the amazing advances in informationalization exemplified by the computer and the internet and advocated by slogans such as 'IT revolution' appear to promise the arrival of bright future. The rapid rises in stock prices of firms related to the information technology seem to endorse these prospects. It seems that all people of the world have now acquired a means to communicate with all the other people at any place. Although there is still the problem of data divide, it seems as if the human race will eventually obtain a means of worldwide solidarity with the development of global networks. Certainly, the situation may allow speculation about the possibility of the development of new communities and individuality based on progresses informationalization. In a sense the new media might seem to restore human communal nature positively.

However, many of the new media that have appeared in rapid succession with informationalization appear to expand the world of virtual reality with a consequent reduction in contact with nature and dilution of the 'body' of the human subject that manipulates the media. Progresses in informationalization make the information environment and cyberspace proliferate and enlarge. Not a few people feel serious worries and threat about the future of children who grow up in such an informationalized environment. Therefore, the problems of informationalization also joins 'environmental problems'. 'Environmental problems' are, thus, closely related to strictly contemporary problems of human beings and their society such as 'life' and 'information'. This must not be overlooked.

Along with the above points, it must not be forgotten that nature that surrounds us is socialized and artificially controlled. If so, the natural environment and social and cultural environment are fused and are in continuity to a large extent, not being clearly distinguishable.

From this viewpoint, evaluation of 'environmental problems' inevitably entails evaluation also of the social and cultural environment. Therefore, 'environmental problems' are profoundly related to problems of not only the natural environment of human beings but also their social and cultural environment. It must be remembered that the social and cultural environment as well as the natural environment is essential to humans, because human beings are innately 'social' and 'conscious' beings as well as 'natural' beings.

Humans cannot become humans even biologically without their unique social and cultural environment. Not only the natural environment but also social and cultural environment is involved intrinsically in the development of man as a biological existence itself. Neonates are not only in their unique relationship with nature but cannot grow into human beings even biologically without relating themselves to their social and cultural environment by communication with their care-givers. As Adolf Portman observed, human babies are born in a premature state compared with other primates. This means that the social and cultural environment is decisively incorporated in the development of infants. However, the understanding that the social and cultural environment largely restricts the biological existence of human beings must not lead to disregard of their biological and natural existence. Inversely, we must find a perspective to radically address the present social and cultural environment of human beings with emphasis on their biological and natural existence.

The relationship between the environment and human beings must be evaluated from diverse viewpoints with the above discussion on 'environmental problems' in mind (refer to *Philosophical Anthropology of Environment and Information: For a Symbiotic and Communal Society* by the author, Aoki Shoten, 2000 for further reading), but we will approach this problem here according to the proposals by Hideo Obara, a unique zoologist and philosopher.

3.2 'Naturalness' in human being

In my former evaluation on cruel 'bullying', I suggested that there are problems with the social, cultural, and mental environment of children in contemporary Japan as a background of bullying and further touched on its relationship with the natural environment (*Communication Today and Symbiosis/Cooperation*, by the author, Aoki Shoten, 1996). One of the points that I raised in this work was that children of contemporary Japan are losing the ability to sympathize with other human beings and living things. In connection, a mention about the approaches of Hideo Obara to 'naturalness in humans' is considered to be relevant. Obara observed as follows.

> In questioning the naturalness in humans, it must be assumed that the state of human existence (the presence and the way in which humans is present) is not simply natural. Still, we question the naturalness of human being, because humans is not entirely a natural being but is a social and cultural being that contains natural parts (In Quest for 'the Naturalness' in Humans, Bulletin of the Kagawa Nutrition College, 27, 1996, pp. 111-112).

Therefore, we must first note that the quest for 'the naturalness in humans' made on such an assumption is on a different level from the argument of 'returning to nature' by simple ecologists. Obara considered that man has modified his own characteristics as he has adapted himself to nature socialized by man and to the artificial ecosystem and called this phenomenon 'self-domestication', using the anthropological knowledge. However, as the expression 'self-domestication' may invite to unintended associations, Obara himself has also used the term, 'socialization in socialized nature'. In essence, these terms are impressive expressions of the conscious and social being unique to man in which, being part of nature himself, socializes nature for himself and modifies himself by interacting with this socialized nature rather than simply interacting with nature that confronts man.

When Ohara questions 'the naturalness in humans' from such a viewpoint, he is aware of problems with the 'environment of children' created by the contemporary society, which is called an advanced consumer society and a highly informationalized society.

> Problems concerning 'the naturalness' in the modern society that are seriously aggravating are those related to children. When the author had a chance to talk to (then) Chairman Takashi Ohta of the Japan Society of Pedagogics, the author realized that naturalness for humans (and its clarification) is needed in basic concepts in the life of children such as play, discipline, and health. It would be obvious that naturalness for a human being (man) is necessary in the environment of children for their development as human beings. For, its realization is considered to be reflected in educational activities in a broad sense [*ibid.*, p. 114].

With this understanding, 'environmental problems' may be regarded as an acute question posed to not only children but also us all about 'the naturalness in humans'. The 'naturalness' questioned here does not simply mean the biological being of man but is a 'Idee' and a value in the dimension that 'transcends' biological limitations of humans similarly to truth, goodness, and beauty. In this sense, the mankind has discovered and sought for concepts and values including truth, goodness, and beauty or freedom, equality, and fellowship through its history. In the modern times, however, environmental problems seem to be proposing that man must pursue the new 'Idea' or value of 'naturalness'.

As is well known, Descartes, the father of modern philosophy, proposed the concept of mechanical nature to replace the concept of teleological nature of Aristotle and considered animals automatic machines. Furthermore, La Mettrie presented the theory of human machine using the clock as a model. The inclination to mechanism that occurred in the modern times has also permeated deep into the modern biology, and the actual living things and life are regarded ultimately as information machinery of DNA as the molecular biologist Jacques Monod described in his best-seller '*Le hasard et la nécessité*' (*The Chance and Necessity*). Mechanism is further expanded and deepened today as the softer and more delicate information mechanism represented by the computer. 'Artificial intelligence' and 'artificial life' often appear as headlines, and robots have gained extreme popularity as substitutes for pets. The 'environment for children' appears to be becoming more and more mechanical in physical and informatic senses and isolated from the sources of nature, turning into a capsule.

What, then, should be done to restore the 'naturalness' of children as human beings in the modern times? There can be diverse approaches including exposing them more to nature and practicing environmental education, but I would like to emphasize here the importance of the experience of communicating with nature. Various measures to have children interact with nature (ex. animals) will induce 'naturalness' in children and provide opportunities for them to experience and consider it by themselves. It will further have them nurture 'communicative attitudes to nature'.

Ironically speaking, Japanese have such attitudes in their traditional culture. The attitudes might not be rational from viewpoint of modern science, but I do not think we can say that these are anti-scientific.

3.3 Grounds for 'communicative attitudes to nature' and 'Kyosei'

To me, it seems important for modern man to restore and develop 'communicative attitudes toward nature', and the following are possible grounds for my argument.

We may be able to take communicative attitudes toward species remotest from us as well as those closest to us in the tree of evolution or living things other than animals and even inanimate natural objects although actual communication with them may not be possible. What is important in this is that such communicative attitudes are not simply fictitious. When we depart from industrial and scientific concepts in which we completely objectify nature on the basis of the mechanical view of nature and regard it simply as resources for economic growth, nature itself is expected to act as a factor that induces communicative attitudes in us in our interaction with nature. This is no mysticism of nature.

We human beings were born in the continuity of evolution by nature itself, and animals akin to man in the tree of evolution are different in species but can be partners of our daily communication. This close communication may be regarded as the core of the concentric circles of communicative attitudes, which become gradually sparse as the circle is more distant from the core, just as the ripples caused by a stone thrown into a pond gradually dwindle as they spread across the pond.

What is important in connection with this is the assertion of the 'subjectivity' of living things by Kinji Imanishi, who led the world by establishing the new methodology of primatology. I think he could introduce 'personification' and 'analogy' as a kind of scientific method into primatology, as he believed in the significance of 'communicative attitudes toward nature' in the Japanese cultures. Imanishi emphasized that living things exercise 'subjectivity' against the environment or other living things in their own ways even though their 'subjectivity' is not as notable as humans, who assert their 'subjectivity' to such a degree as to modify nature. The problem is weakening of this subjectivity intrinsic to man as a species as he expands his unique conscious and cultural subjectivity. Restoration of the subjectivity of human being as a living thing will be important in connection with communicative attitudes to nature. (*The World of Living Things*, [1941] by Kinji Imanishi, had an enormous impact in Japan, both on intellectuals and on the general public, very little is known about it in the English-speaking world, but recently it was translated into the English).

Now we can return to the problem with 'symbiosis between humans and nature'. We need such communicative attitudes to nature for realizing 'symbiosis between humans and nature', which we understand.

Such communicative attitudes to nature lead to respect for nature, but this does not mean expansion of objects of rights as described by advocates of 'rights of nature' rather than understanding nature as 'the subject of rights'. Loss of rich nature as a partner of communication or loss of communicative attitudes to it may not directly endanger man's survival on the level of clothing, diet, and housing. However, in consideration of the fact that some aspects of man's mind have been formed on the basis of such communicative relations with nature in the long history of mankind, and in consideration of the fact that man is essentially a 'communicative being' from 'communal being', which I emphasized before, such losses are expected to bring about crises to the survival of humans in a broad sense.

Although nature-centered contentions ('physiocentrism' against 'anthropocentrism') of deep ecology cannot be supported as they are, I sympathize with them in that they reflect more serious crises of this kind. However, as observed in various natural disasters, the fact that nature can literally be 'ferocious' must not be forgotten. Therefore, while the view of 'conquest of nature' must be overcome, problems of 'controlling nature' and rational intervention in nature are not expected to vanish in 'symbiosis between humans and nature' as we understand it. Thus, I consider that the concept of 'symbiosis between humans and nature' may be properly understood pursuing it from the viewpoint of 'communication', which will be internally associated with the sustainable society.

The argument that exploitation of nature by humans is immediately 'unnatural' cannot be supported. Humans, who are desire-bound beings, have no choice but to utilize nature as the object of his desires. The problem is what desires we should seek to fulfill. In the modern social systems, there is always fundamental drive for economic 'growth', which is expressed through human existence as ever swelling desire for money and commodities. That is, our society fosters the fundamental 'anti-natural' drive to inflate desire to consume and to over-exploit nature for incremental regeneration and self-proliferation of systems.

Therefore, the true confrontation between humans and nature in the contemporary times is not that of 'anthropocentrism' *versus* 'physiocentrism' based on 'juxtaposition of humans and nature', which has often been observed in environmental ethics, but it is confrontation of 'organic nature and humans' *versus* the 'self-driven socioeconomic system'. The latter is a system that was produced by humans but ironically continues to move and expand by itself for self-proliferation by utilizing nature and humans as its means and resources.

Therefore, I insist that the concept 'symbiosis between humans and nature' as I understand it, will transcend both 'anthropocentrism' and 'physiocentrism'. The philosophical Idee 'kyosei' will urge the mankind to seek for 'communicative attitudes to nature' on the basis of the concept of 'naturalness' of humans and to restore 'communal nature' of human beings through solidaristic control of the 'self-driven socioeconomic system'. Also, that will gradually change its social systems into sustainable society.

Bibliography

P. J. Asquith (ed.), *Japanese View of Nature: The World of Living Things*, by Kinji Imanishi, Routledge, London and NewYork 2002.

K. Kurokawa, *The Philosophy of Symbiosis: Lifestyle for survival in the future*, Tokuma Shoten, Tokyo 1987.

K. Imanishi, *The World of Living Things*, Koudansya, Tokyo 1993.

A. Ness, *Ecology, Community and Lifestyle*, Cambridge University Press, Cambridge 1989.

H. Obara, *Metamorphosis of Homo sapiens today*, Asahi Shinbun Publishing, Tokyo 2000.

S. Ozeki, *Japanese View of Nature and Environmental Philosophy Today: Kinji Imanisi' View of Nature, in Contemporary Significance of the Eastern Thought*, Nobunkyo, Tokyo 1999.

S. Ozeki, *Communication Today and Symbiosis/Cooperation*, Aoki Shoten Publishing Co. Ltd., Tokyo 2000.
S. Ozeki, *Philosophical Antholopology of Environment and Information: For a Symbiotic and Communal Society*, Aoki Shoten Publishing Co. Ltd., Tokyo 2000.
T. Umehara, *Philosophy of Symbiosis and Circulation*, Syogakkan, Tokyo 1996.

Shuji Ozeki, Department of Ecoregion Science, Tokyo University of Agriculture and Technology

Yearbook of the Artificial

Published volumes

Vol. 1 Yearbook of the Artificial: Methodological Aspects and Cultural Issues, 272 pages, 2002.
Vol. 2 Yearbook of the Artificial: Models in Contemporary Sciences, 379 pages, 2004.
Vol. 3 Yearbook of the Artificial: Cultural Dimensions of the User, 249 pages, 2005.
Vol. 4 Yearbook of the Artificial: *Kyosei,* Culture and Sustainable Technology, 216 pages, 2006.

FSC
www.fsc.org
MIX
Papier | Fördert gute Waldnutzung
FSC® C083411

Kolibri 360 GmbH
Ferdinand-Jühlke-Straße 7
99095 Erfurt, Deutschland
produktsicherheit@kolibri360.de

Druck:
CPI Druckdienstleistungen GmbH
im Auftrag der
Kolibri 360 GmbH
Ferdinand-Jühlke-Str. 7
99095 Erfurt